I0489307

September 2012

UNCONVENTIONAL OIL AND GAS DEVELOPMENT

Key Environmental and Public Health Requirements

GAO

Accountability ★ Integrity ★ Reliability

GAO-12-874

GAO
Accountability * Integrity * Reliability

Highlights

Highlights of GAO-12-874, a report to congressional requesters

UNCONVENTIONAL OIL AND GAS DEVELOPMENT

Key Environmental and Public Health Requirements

Why GAO Did This Study

Technological improvements have allowed the extraction of oil and natural gas from onshore unconventional reservoirs such as shale, tight sandstone, and coalbed methane formations. Specifically, advances in horizontal drilling techniques combined with hydraulic fracturing (pumping water, sand, and chemicals into wells to fracture underground rock formations and allow oil or gas to flow) have increased domestic development of oil and natural gas from these unconventional reservoirs. The increase in such development has raised concerns about potential environmental and public health effects and whether existing federal and state environmental and public health requirements are adequate.

GAO was asked to review environmental and public health requirements for unconventional oil and gas development and (1) describe federal requirements; (2) describe state requirements; (3) describe additional requirements that apply on federal lands; and (4) identify challenges, if any, that federal and state agencies reported facing in regulating oil and gas development from unconventional reservoirs. GAO identified and analyzed federal laws, state laws in six selected states (Colorado, North Dakota, Ohio, Pennsylvania, Texas, and Wyoming), and interviewed federal and state officials and representatives from industry, environmental, and public health organizations.

GAO is not making recommendations. In commenting on the report, agencies provided information on recent regulatory activities and technical comments.

View GAO-12-874. For more information, contact David C. Trimble at (202) 512-3841 or trimbled@gao.gov.

What GAO Found

As with conventional oil and gas development, requirements from eight federal environmental and public health laws apply to unconventional oil and gas development. For example, the Clean Water Act (CWA) regulates discharges of pollutants into surface waters. Among other things, CWA requires oil and gas well site operators to obtain permits for discharges of produced water—which includes fluids used for hydraulic fracturing, as well as water that occurs naturally in oil- or gas-bearing formations—to surface waters. In addition, the Resource Conservation and Recovery Act (RCRA) governs the management and disposal of hazardous wastes, among other things. However, key exemptions or limitations in regulatory coverage affect the applicability of six of these environmental and public health laws. For example, CWA also generally regulates stormwater discharges by requiring that facilities associated with industrial and construction activities get permits, but the law and its regulations largely exempt oil and gas well sites. In addition, oil and gas exploration and production wastes are exempt from RCRA hazardous waste requirements based on a regulatory determination made by the Environmental Protection Agency (EPA) in 1988. EPA generally retains its authorities under federal environmental and public health laws to respond to environmental contamination.

All six states in GAO's review implement additional requirements governing activities associated with oil and gas development and have updated some aspects of their requirements in recent years. For example, all six states have requirements related to how wells are to be drilled and how casing—steel pipe within the well—is to be installed and cemented in place, though the specifics of their requirements vary. The states also have requirements related to well site selection and preparation, which may include baseline testing of water wells before drilling or stormwater management.

Oil and gas development on federal lands must comply with applicable federal environmental and state laws, as well as additional requirements. These requirements are the same for conventional and unconventional oil and gas development. The Bureau of Land Management (BLM) oversees oil and gas development on approximately 700 million subsurface acres. BLM regulations for leases and permits govern similar types of activities as state requirements, such as requirements for how operators drill the well and install casing. BLM recently proposed new regulations for hydraulic fracturing of wells on public lands.

Federal and state agencies reported several challenges in regulating oil and gas development from unconventional reservoirs. EPA officials reported that conducting inspection and enforcement activities and having limited legal authorities are challenges. For example, conducting inspection and enforcement activities is challenging due to limited information, such as data on groundwater quality prior to drilling. EPA officials also said that the exclusion of exploration and production waste from hazardous waste regulations under RCRA significantly limits EPA's role in regulating these wastes. In addition, BLM and state officials reported that hiring and retaining staff and educating the public are challenges. For example, officials from several states and BLM said that retaining employees is difficult because qualified staff are frequently offered more money for private sector positions within the oil and gas industry.

_____ **United States Government Accountability Office**

Contents

Letter		1
	Background	5
	Federal Environmental and Public Health Laws Apply to Unconventional Oil and Gas Development but with Key Exemptions	17
	States in Our Review Implement Additional Requirements and Recently Updated Some Requirements	47
	Additional Requirements Apply on Federal Lands	68
	Federal and State Agencies Reported Several Challenges Regulating Unconventional Oil and Gas Development	77
	Agency Comments and Our Evaluation	81
Appendix I	Objectives, Scope, and Methodology	84
Appendix II	Key Requirements and Authorities under the Safe Drinking Water Act	88
	Underground Injection Control Program	88
	Imminent and Substantial Endangerment Authorities	97
Appendix III	Key Requirements and Authorities under the Clean Water Act	99
	National Pollutant Discharge Elimination System Program	100
	Oil and Hazardous Substances Spill Prevention, Reporting, and Response	117
Appendix IV	Key Requirements and Authorities under the Clean Air Act	126
	National Emission Standards for Hazardous Air Pollutants	129
	New Source Performance Standards	137
	New Source Review	142
	Title V Operating Permits	143
	Source Determinations and Aggregation Issues for Title V and NSR	145
	Greenhouse Gas Reporting Rule	147
	Accidental Releases	149
	EPA Enforcement Authorities	155
	Imminent and Substantial Endangerment Authority	155

Appendix V	Key Requirements and Authorities under the Resource Conservation and Recovery Act	156
	Subtitle C – Hazardous Waste	156
	Subtitle D – Solid Waste	165
	Enforcement	166
	Imminent and Substantial Endangerment Authority	168
Appendix VI	Key Requirements and Authorities under the Comprehensive Environmental Response, Compensation, and Liability Act	171
Appendix VII	Key Requirements and Authorities under the Emergency Planning and Community Right-to-Know Act	179
Appendix VIII	Key Requirements and Authorities under the Toxic Substances Control Act	188
Appendix IX	Selected State Requirements	192
Appendix X	Crosswalk between Selected Requirements from EPA, States, and Federal Lands	225
Appendix XI	Comments from the Department of Agriculture	229
Appendix XII	Comments from the Department of the Interior	230
Appendix XIII	GAO Contact and Staff Acknowledgments	233

Tables

Table 1: Potential Waste Management and Disposal Options[a]   14

Table 2: Exemptions or Limitations in Regulatory Coverage for the Oil and Gas Exploration and Production Industry in Six Environmental Laws   44

Table 3: Key EPA Response Authorities Relevant to Oil and Gas Well Sites   46

Table 4: Key Federal Environmental and Public Health Requirements and State Requirements for Oil and Gas Production Wells   48

Table 5: Chemical Disclosure Requirements in Six Selected States   54

Table 6: Surface Agency Roles in Leasing and Permitting Federal Minerals   70

Table 7: Agencies and Organizations Contacted   85

Table 8: Summary of Effluent Limitations Guidelines for Wastewater Discharges from Selected Subcategories of Oil and Gas Wells Located on Land   103

Table 9: Summary of CAA Programs That May Apply to Emissions from Oil and Gas Well Sites   128

Table 10: NSPS for Natural Gas Wells, by Well Subcategory   139

Table 11: Primary State Agencies Responsible for Regulating Oil and Gas Development in Six States   192

Table 12: Selected State Requirements—Siting and Site Preparation   193

Table 13: Selected State Requirements—Drilling, Casing, and Cementing   196

Table 14: Selected State Requirements—Hydraulic Fracturing   202

Table 15: Selected State Requirements—Well Plugging   208

Table 16: Selected State Requirements—Site Reclamation   209

Table 17: Selected State Requirements—Waste Management in Pits   211

Table 18: Selected State Requirements—Waste Management through Underground Injection   216

Table 19: Selected State Requirements—Managing Air Emissions   222

Table 20: Crosswalk between Selected Requirements from EPA, Six States, and Federal Minerals   225

Figures

Figure 1: Conventional and Unconventional Oil and Gas Reservoirs   6

Figure 2: Locations of Unconventional Reservoirs in the United States   7

Figure 3: Well Pad and Freshwater Storage Tanks   9

Figure 4: Horizontal Drilling and Hydraulic Fracturing in an
      Unconventional Shale Formation   11
Figure 5: Hydraulic Fracturing in a Coalbed Methane Formation   12
Figure 6: Potential Sources and Types of Air Emissions from Oil
      and Gas Development   16
Figure 7: Enhanced Recovery and Disposal Wells   19

Abbreviations

APD	application for permit to drill
BLM	Bureau of Land Management
BTEX	benzene, toluene, ethylbenzene, xylenes
CAA	Clean Air Act
CAS	Chemical Abstracts Service
CERCLA	Comprehensive Environmental Response, Compensation, and Liability Act
CWA	Clean Water Act
EPA	Environmental Protection Agency
EPCRA	Emergency Planning and Community Right-to-Know Act
FIFRA	Federal Insecticide, Fungicide, and Rodenticide Act
FWS	Fish and Wildlife Service
H_2S	hydrogen sulfide
HAP	hazardous air pollutant
MACT	maximum achievable control technology
NAICS	North American Industry Classification System
NEPA	National Environmental Policy Act
NESHAP	National Emissions Standards for Hazardous Air Pollutants
NORM	naturally-occurring radioactive material
NPDES	National Pollutant Discharge Elimination System
NSPS	New Source Performance Standards
NSR	New Source Review
OEPA	Ohio Environmental Protection Agency
POTW	publicly-owned treatment works
PSD	Prevention of Significant Deterioration
psi	pounds per square inch
RCRA	Resource Conservation and Recovery Act
SDWA	Safe Drinking Water Act
SIP	state implementation plan
SPCC	Spill Prevention, Control, and Countermeasure
STRONGER	State Review of Oil and Natural Gas Environmental Regulations
TRI	Toxics Release Inventory
TSCA	Toxic Substances Control Act
UIC	Underground Injection Control
VOC	volatile organic compound

This is a work of the U.S. government and is not subject to copyright protection in the United States. The published product may be reproduced and distributed in its entirety without further permission from GAO. However, because this work may contain copyrighted images or other material, permission from the copyright holder may be necessary if you wish to reproduce this material separately.

United States Government Accountability Office
Washington, DC 20548

September 5, 2012

Congressional Requesters

For decades, the United States has imported oil and natural gas to fuel vehicles and to heat and power homes and businesses. However, improvements in technology have allowed companies that develop petroleum resources to extract oil and natural gas from onshore unconventional reservoirs that were previously considered inaccessible because traditional techniques did not yield sufficient amounts for economically viable production. For purposes of this report, unconventional reservoirs include shale, tight sandstone,[1] and coalbed methane formations. Specifically, advances in horizontal drilling techniques combined with hydraulic fracturing have recently increased domestic production of oil and natural gas from such onshore unconventional reservoirs. These advances, which have taken place over the last several decades, now allow operators—companies that extract oil and natural gas—to accurately determine the location of a drill bit while drilling thousands of feet horizontally. Hydraulic fracturing involves pumping water, sand, and chemical additives into oil and gas wells at high enough pressure to fracture underground rock formations and allow oil or gas to flow. When combined with horizontal drilling, hydraulic fracturing allows operators to fracture the rock formation along the entire horizontal portion of a well, increasing the number of pathways through which oil or gas can flow. According to the Energy Information Administration,[2] oil production from shale formations (shale oil)[3] has increased significantly in several areas of the country, including the

---

[1]Conventional sandstone has well-connected pores, but tight sandstone has irregularly distributed and poorly connected pores. Due to this low connectivity or permeability, gas trapped within tight sandstones is not easily produced.

[2]The Energy Information Administration is the statistical and analytical agency within the Department of Energy that collects, analyzes, and disseminates independent and impartial information on energy issues.

[3]Shale oil differs from "oil shale." Oil shale requires a different process to extract. Specifically, to extract the oil from oil shale, the rock needs to be heated to very high temperatures—ranging from about 650 to 1,000 degrees Fahrenheit—in a process known as retorting. For additional information on oil shale, see GAO, *Energy-Water Nexus: A Better and Coordinated Understanding of Water Resources Could Help Mitigate the Impacts of Potential Oil Shale Development*, GAO-11-35 (Washington, D.C.: Oct. 29, 2010).

Bakken formation in North Dakota and Montana, where production increased from just over 2,000 barrels per day in 2000 to approximately 500,000 barrels per day in 2012. The Energy Information Administration also projects that natural gas production from shale (shale gas) will account for almost half of domestic production by 2035. According to a 2011 report by the Department of Energy, the recent substantial growth in domestic natural gas production from shale has already brought lower natural gas prices, domestic jobs, and the prospect of enhanced national security.[4]

However, the increase in oil and gas development from unconventional reservoirs has raised concerns about the potential environmental and public health effects of such development.[5] For example, several environmental groups have expressed concerns that this development releases hazardous air pollutants, such as benzene, and may contaminate underground drinking water supplies and surface waters due to spills or faulty well construction. These concerns have raised questions about existing federal and state requirements governing oil and gas development from unconventional reservoirs on both private and federal lands. The Environmental Protection Agency (EPA) administers and enforces key federal laws, such as the Safe Drinking Water Act, the Clean Water Act, and others, that aim to protect human health and the environment. Under these statutes, EPA and its Regional offices work with states that implement aspects of some of these laws, as well as additional state requirements. EPA is also conducting a study, as directed by a congressional committee, to examine the potential effects of hydraulic fracturing on drinking water resources.[6] In addition, federal land management agencies, including the Department of the Interior's Bureau of Land Management (BLM), National Park Service, Fish and Wildlife Service (FWS) and the U.S. Department of Agriculture's Forest Service

---

[4]Department of Energy, Secretary of Energy Advisory Board, *Shale Gas Production Subcommittee 90-Day Report* (Washington, D.C.: 2011)

[5]GAO is conducting a separate and more detailed review of risks associated with shale oil and gas development.

[6]EPA announced in March 2010 that its Office of Research and Development would conduct a study to examine the potential effects of hydraulic fracturing on drinking water resources. According to EPA officials, the agency anticipates issuing a progress report in 2012 and a final report in 2014.

manage federal lands and have responsibilities for oil and gas development.[7]

You asked us to review environmental and public health requirements for oil and gas development from onshore unconventional reservoirs. For such development, this report (1) describes federal environmental and public health requirements; (2) describes state requirements; (3) describes additional requirements that apply on federal lands; and (4) identifies challenges, if any, that federal and state agencies reported facing in regulating oil and gas development from unconventional reservoirs.

To identify federal environmental and public health requirements governing onshore oil and gas development from unconventional reservoirs, we identified and analyzed eight key federal environmental and public health laws and corresponding regulations and guidance. The eight federal environmental and public health laws are: the Safe Drinking Water Act (SDWA); Clean Water Act (CWA); Clean Air Act (CAA); Resource Conservation and Recovery Act (RCRA); Comprehensive Environmental Response, Compensation, and Liability Act (CERCLA); Emergency Planning and Community Right-to-Know Act (EPCRA); Toxic Substances Control Act (TSCA); and Federal Insecticide, Fungicide, and Rodenticide Act (FIFRA). We focused our analysis on federal and state requirements that apply to activities on the well site—the area of land where drilling takes place—and wastes or emissions generated at the well site rather than on downstream infrastructure such as pipelines or refineries. We also analyzed state laws and regulations in a nonprobability sample of six selected states—Colorado, North Dakota, Ohio, Pennsylvania, Texas, and Wyoming. We selected states with current unconventional oil or gas development, as well as those states with large reservoirs of unconventional oil or gas, a variety of types of unconventional reservoirs, differing historical experiences with the oil and gas industry, and significant development on federal lands. Because we used a nonprobability sample, the information that we collected from these states cannot be generalized to all states but can provide illustrative examples. We also reviewed laws, regulations, and guidance governing oil and gas development on federal lands. In addition, we reviewed

---

[7]BLM also manages oil and gas development on American Indian lands, but American Indian lands are outside the scope of this review.

several reports issued by environmental and public health organizations, industry, academic institutions, and government agencies that provided perspectives on federal and state regulations.

To complement our analysis of laws and regulations, we interviewed officials in EPA headquarters and the four Regional offices responsible for overseeing implementation of federal programs in the six selected states; oil and gas and environmental regulators for the six selected states; officials in BLM headquarters and field offices in each of the three selected states with significant amounts of federal land; and officials in Park Service, Forest Service, and FWS headquarters to discuss how federal and state requirements apply to the oil and gas industry and any challenges faced by regulators in implementing these requirements. We also contacted representatives from industry, environmental, and public health organizations regarding federal and state regulatory requirements for unconventional oil and gas development. In addition, we met with company officials to discuss federal and state regulations and visited drilling, hydraulic fracturing, and production sites in Pennsylvania and North Dakota. Oil and gas development may also be subject to local laws and requirements related to water use and withdrawals, but we did not include an analysis of these issues in the scope of our review. We also did not include an analysis of the extent to which federal or state laws and regulations concerning oil and gas development apply on tribal lands, or the extent to which tribal laws may apply. In describing federal and state requirements for oil and gas development from unconventional reservoirs, where it is helpful to further the understanding of the requirements, we provide examples of how these requirements have been applied, but these examples do not attempt to provide a comprehensive view of the extent to which enforcement actions have been taken for any of the requirements. A more detailed description of our objectives, scope, and methodology is presented in appendix I.

We conducted this performance audit from November 2011 to September 2012 in accordance with generally accepted government auditing standards. Those standards require that we plan and perform the audit to obtain sufficient, appropriate evidence to provide a reasonable basis for our findings and conclusions based on our audit objectives. We believe that the evidence obtained provides a reasonable basis for our findings and conclusions based on our audit objectives.

Background

Oil and gas reservoirs vary in their geological makeup, location, and size. Regardless of the reservoir, unconventional oil and gas development involves a number of activities, many of which are also conducted in conventional oil and gas drilling. This section describes the types and locations of oil and gas reservoirs and the key stages of oil and gas development.

Types and Locations of Oil and Gas Reservoirs

Oil and natural gas are found in a variety of geologic formations. In conventional reservoirs, oil and gas can flow relatively easily through a series of pores in the rock to the well. Shale and tight sandstone formations generally have low permeability and therefore do not allow oil and gas to easily flow to the well. Shale and tight sandstone formations can occur at varying depths, including thousands of feet beneath the surface. For example, the Bakken shale formation in North Dakota and Montana ranges from 4,500 to 11,000 feet beneath the surface. Coalbed methane formations, often located at shallow depths of several hundred to 3,000 feet, are generally formations through which gas can flow more freely; however, capturing the gas requires operators to pump water out of the coal formation to reduce the pressure and allow the gas to flow. Shale, tight sandstone, and coalbed methane formations are located within basins, which are large-scale geological depressions, often hundreds of miles across, which also may contain other oil and gas resources. There is no clear and consistently agreed upon distinction between conventional and unconventional oil and gas, but unconventional sources generally require more complex and expensive technologies for production, such as the combination of horizontal drilling and multiple hydraulic fractures. See figure 1 for a depiction of conventional and unconventional reservoirs.

Figure 1: Conventional and Unconventional Oil and Gas Reservoirs

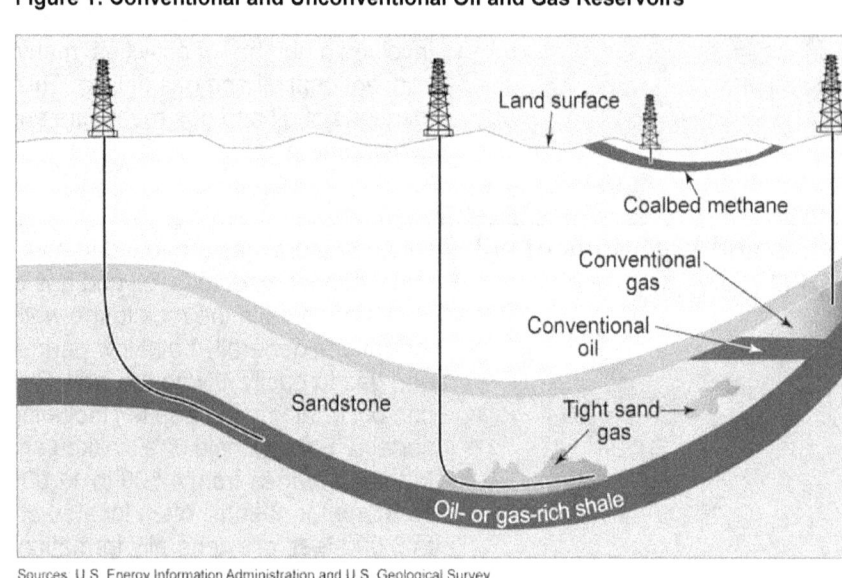

Sources: U.S. Energy Information Administration and U.S. Geological Survey.

Unconventional reservoirs are located throughout the continental United States on both private lands and federal lands that are administered by BLM, Forest Service, Park Service, and FWS (see fig. 2).[8]

---

[8]Unconventional reservoirs may also be located on tribal lands, but these were outside the scope of our review.

Shale basins

Tight sandstone basins

Coalbed methane basins

Source: U.S. Energy Information Administration

Note: The map for tight sandstone basins is based off Energy Information Administration data for "tight gas," which includes both tight sandstone and tight carbonate formations.

Activities Associated With Oil and Gas Development

Developing unconventional reservoirs involves a variety of activities, many of which are also conducted in conventional oil and gas drilling.[9]

Siting and site preparation. The operator identifies a location for the well and prepares the area of land where drilling will take place—referred to as a well pad. In some cases, the operator will build new access roads to transport equipment to the well pad or install new pipelines to transport the oil or gas that is produced. In addition, the operator will clear vegetation from the area and may place storage tanks (also called vessels) or construct pits on the well pad for temporarily storing fluids (see fig. 3). In some cases, multiple wells will be located on a single well pad.

---

[9]Prior to beginning these activities, the operator must acquire the necessary leases to gain the right to drill for oil or gas on a particular area of land. Leasing is not discussed here because it is outside the scope of our review; however, aspects of BLM's process for issuing leases that are relevant to environmental requirements are discussed later in this report. In addition, the operator must obtain a permit to drill from applicable federal or state agencies, but the details of the permitting process are not generally within the scope of our review.

Figure 3: Well Pad and Freshwater Storage Tanks

Source: GAO

Drilling, casing, and cementing. The operator conducts several phases of drilling to install multiple layers of steel pipe—called casing—and cement the casing in place. The layers of steel casing are intended to isolate the internal portion of the well from the outlying geological formations, which may include underground drinking water supplies. As the well is drilled deeper, progressively narrower casing is inserted further down the well and cemented in place. Throughout the drilling process, a special lubricant called drilling fluid, or drilling mud, is circulated down the well to lubricate the drilling assembly and carry drill cuttings (essentially rock fragments created during drilling) back to the surface. After vertical drilling is complete, horizontal drilling is conducted by slowly angling the drill bit until it is drilling horizontally. Horizontal stretches of the well typically range from 2,000 feet to 6,000 feet long but can be as long as 12,000 feet in some cases.

Hydraulic fracturing. Operators sequentially perforate steel casing and pump a fluid mixture down the well and into the target formation at high enough pressure to cause the rock within the target formation to fracture. The sequential fracturing of a well can use between 2 million and 5.6 million gallons of water.[10] Operators add a proppant, such as sand, to the mixture to keep the fractures open despite the large pressure of the overlying rock. About 98 percent of the fluid mixture used in hydraulic fracturing is water and sand, according to a report about shale gas development by the Ground Water Protection Council.[11] The fluid mixture—or hydraulic fracturing fluid—generally contains a number of chemical additives, each of which is designed to serve a particular purpose. For example, operators may use a friction reducer to minimize friction between the fluid and the pipe, acid to help dissolve minerals and initiate cracks in the rock, and a biocide to eliminate bacteria in the water that cause corrosion. The number of chemicals used and their concentrations depend on the particular conditions of the well. After hydraulic fracturing, a mixture of fluids, gases, and dissolved solids flows to the surface (flowback),[12] after which production can begin, and the well is said to have been completed. Operators use hydraulic fracturing in many shale and tight sandstone formations (see fig. 4). Some coalbed methane wells are hydraulically fractured (see fig. 5), but operators may use different combinations of water, sand, and chemicals than with other unconventional wells. In addition, operators must "dewater" coalbed methane formations in order to get the natural gas to begin flowing—a process that can generate large amounts of water.

---

[10]In acquiring this water, operators must comply with applicable state and regional laws or rules regarding water withdrawals, but these are outside the scope of this report.

[11]The Ground Water Protection Council is a nonprofit organization whose members are state groundwater regulatory agencies. Ground Water Protection Council and ALL Consulting. "Modern Shale Gas Development in the United States: A Primer." Prepared for the Department of Energy and National Energy Technology Laboratory. April 2009.

[12]Not all the fluids injected into the well during hydraulic fracturing necessarily flow back to the surface.

Figure 4: Horizontal Drilling and Hydraulic Fracturing in an Unconventional Shale Formation

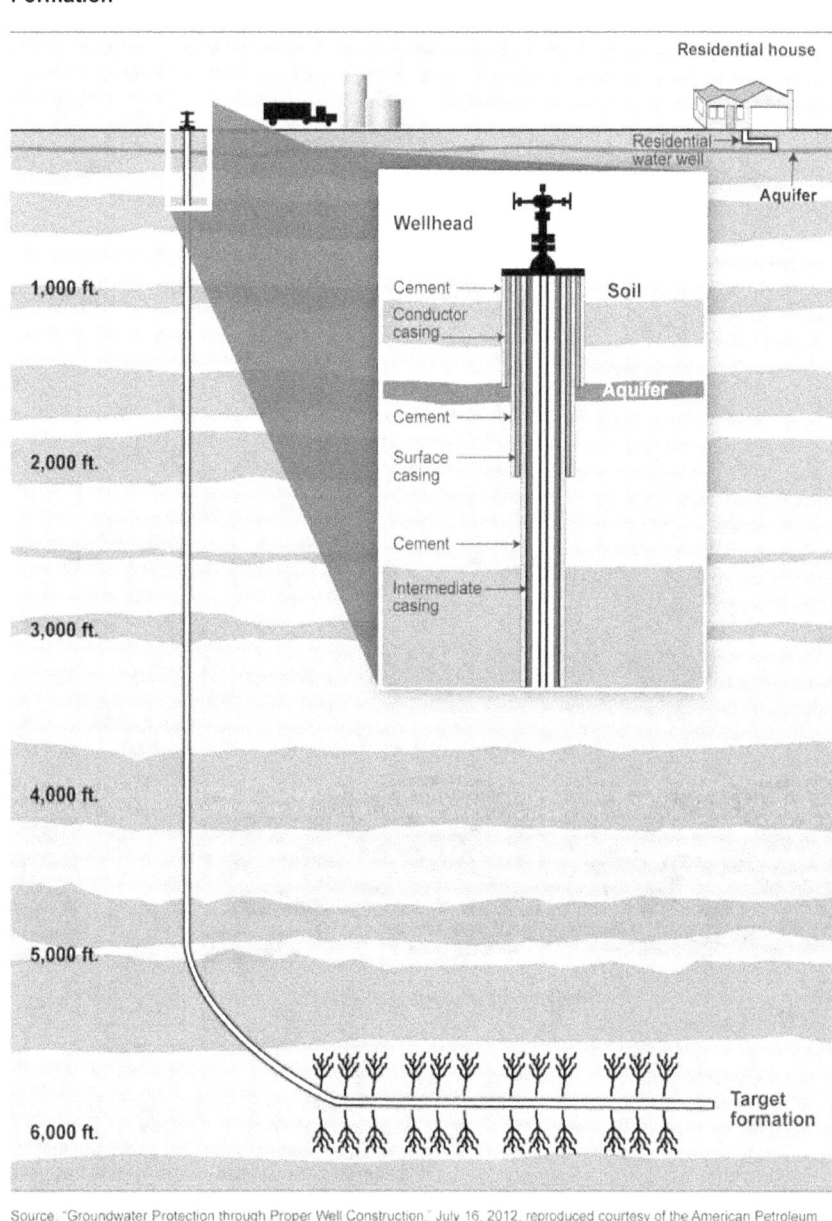

Source: "Groundwater Protection through Proper Well Construction," July 16, 2012, reproduced courtesy of the American Petroleum Institute.

Note: Shale formations can occur at varying depths, including thousands of feet beneath the surface, and according to the American Petroleum Institute, are separated by multiple layers of impervious rock. This figure, which is not to scale, provides a generalized depiction of subsurface geology and well construction; wells may be constructed in a variety of different ways.

Figure 5: Hydraulic Fracturing in a Coalbed Methane Formation

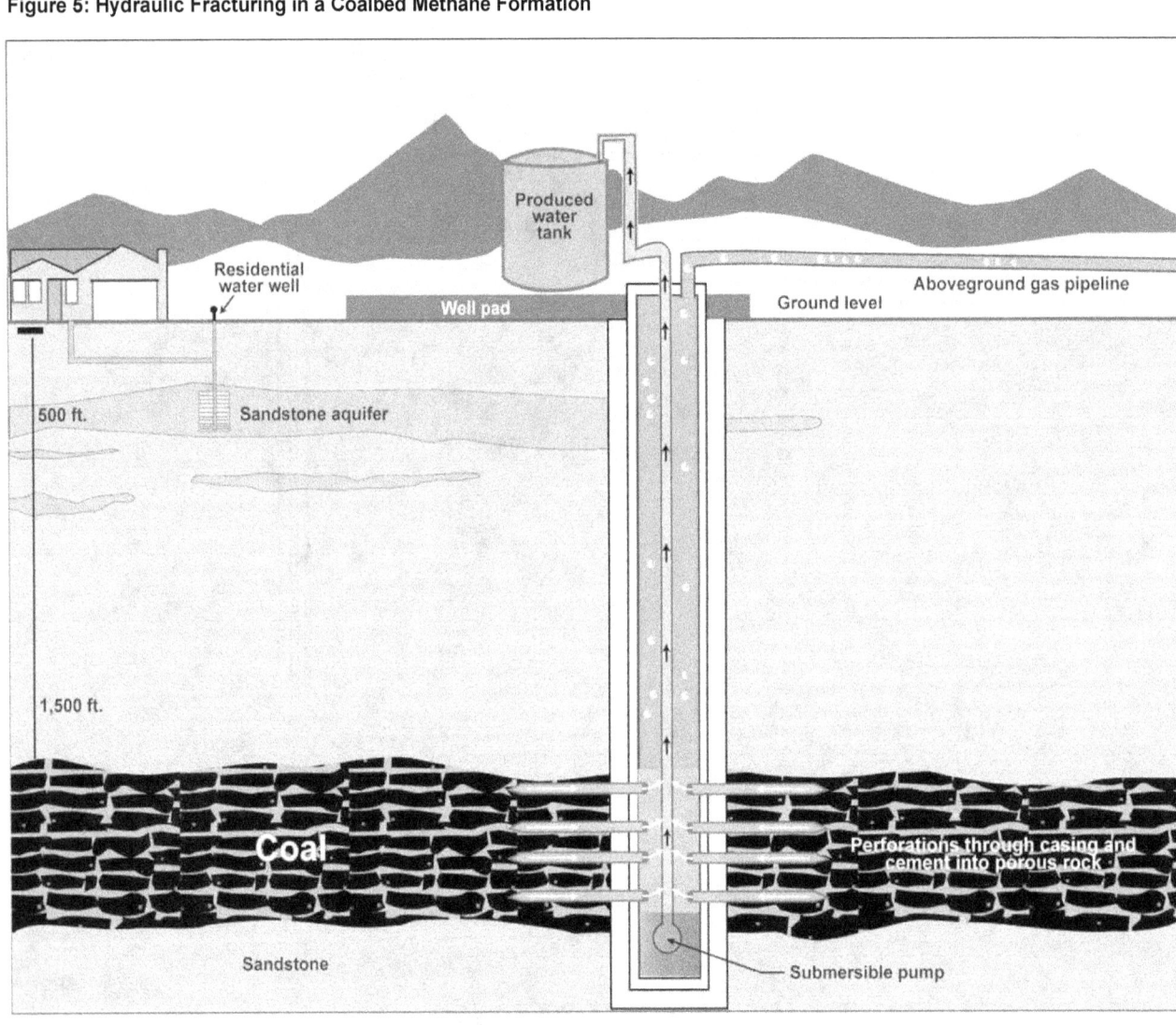

Source: Ecova, Inc.

Note: The water pressure within coalbed methane formations forces natural gas to adhere to the coal. Capturing the gas requires operators to pump water out of the coal formation to reduce the pressure, allowing the natural gas to release (desorb) from the surface of the coal, diffuse through micropores, and then flow through coal cleats (natural fracture systems) into the well. This figure depicts one possible configuration of a coalbed methane well, but other configurations, including aboveground pumps, horizontal drilling, different types of casing and cementing, and deeper or shallower coalbed methane formations are also possible.

GAO-12-874 Unconventional Oil and Gas Development

Well plugging. Once a well is no longer producing economically, the operator typically plugs the well with cement to prevent fluid migration from outlying formations into the well and to prevent downward drainage from inside the well. In some cases, wells may be temporarily plugged so that the operator has the option of reopening the well in the future. In some states with a long history of oil and gas development, wells drilled decades ago may not have been properly plugged—or the plug may have deteriorated.

Site reclamation. Once the well is plugged, the operator takes steps to restore the site to make it acceptable for specific uses, such as farming. For example, reclamation may involve removing equipment from the well pad, closing pits, backfilling soil, and restoring vegetation.[13] Sometimes, when a well starts production, operators reclaim the portions of a site affected by the initial drilling activity.

Waste management and disposal. Throughout the drilling, hydraulic fracturing, and subsequent production activities, operators must manage and dispose of several types of waste. For example, operators must manage produced water, which, for purposes of this report includes flowback water—the water, proppant, and chemicals used for hydraulic fracturing—as well as water that occurs naturally in the oil- or gas-bearing geological formation. Operators temporarily store produced water in tanks or pits, and some operators may recycle it for reuse in subsequent hydraulic fracturing. Options for permanently disposing of produced water vary and may include, for example, injecting it underground into wells designated for such purposes.[14] Operators also generate solid wastes such as drill cuttings and could potentially generate small quantities of hazardous waste. See table 1 for additional methods for managing and disposing of waste.

---

[13]Backfilling is refilling a pit or other area with soil.

[14]We recently issued a report on the quantity, quality, and management of water produced during oil and gas production. See GAO, *Energy-Water Nexus: Information on the Quantity, Quality, and Management of Water Produced during Oil and Gas Production*, GAO-12-156 (Washington, D.C.: Jan. 9, 2012).

Table 1: Potential Waste Management and Disposal Options[a]

	Liquid waste[b]	Solid waste	Hazardous waste
Primary types of waste	• Produced water • Drilling mud	• Drill cuttings • Trash	• Unused hydraulic fracturing chemicals being discarded • Certain other chemical and oily wastes
Options for temporary storage	• Tanks or pits	• Tanks or pits	• Tanks
Options for reuse	• Recycle for use in future hydraulic fracturing • Irrigation • Roadspreading (used for dust or ice suppression) • Reuse of drilling mud	• Roadspreading of drill cuttings	N/A
Options for permanent disposal	• Underground injection well • Discharge to surface water • Commercial treatment facilities • Publicly-owned treatment works	• Solid waste landfill • Bury drill cuttings on or near well pad	• Hazardous waste disposal facility

Source: GAO.

[a]This table identifies a range of temporary storage and permanent disposal options but may not include all available options. Depending on the region or state, some practices may not be technically feasible or legally permissible. The table lists potential disposal options; in some cases, treatment may be required before disposal.

[b]Liquid wastes may be considered solid or hazardous wastes under applicable law. Liquid wastes are listed separately here because liquid wastes may be managed differently from wastes that are more solid in form.

Managing air emissions. Throughout the drilling, hydraulic fracturing, and production activities, operators also are to manage air emissions. There are four key types of air emissions that may occur at oil and gas well sites as follows:

- Criteria pollutants are a set of common air pollutants that include ground level ozone, nitrogen oxides, particulate matter, sulfur dioxide, and carbon monoxide.[15] Ground level ozone is created by chemical reactions between nitrogen oxides and volatile organic compounds and is associated with a wide range of adverse health effects that range from decreased lung function to hospital admissions for

---

[15]The other criteria pollutant is lead, but it is not commonly associated with oil and gas development.

respiratory causes.[16] Particulate matter is a complex mixture of small particles and liquid droplets, which are linked to a variety of health problems such as cardiovascular and respiratory problems. Nitrogen oxides have been linked to respiratory illness. Short-term exposure to sulfur dioxide is linked to a number of adverse respiratory effects, including the narrowing of airways and increased asthma symptoms. Carbon monoxide can cause harmful health effects by reducing oxygen delivery to the body's organs (like the heart and brain).

- Hazardous air pollutants, such as benzene, are pollutants known or suspected to cause cancer or other serious health effects, such as birth defects, or adverse environmental effects.

- Hydrogen sulfide is a toxic and flammable gas that poses safety and health hazards to workers at the well site.

- Methane is a greenhouse gas that, according to some estimates, is over 20 times more efficient in trapping heat in the atmosphere than carbon dioxide—another greenhouse gas—over a 100-year period.

Emissions related to oil and gas production are from both stationary sources and mobile sources (see fig. 6). Stationary sources include wells, pumps, storage vessels, pneumatic controllers, dehydrators, pits, and flaring.[17] Mobile sources include trucks bringing fuel, water, or supplies to the well site; construction vehicles; and some truck-mounted pumps or engines used for drilling or hydraulic fracturing.

---

[16]Volatile organic compounds are emitted as gases from certain solids or liquids and include a variety of chemicals, some of which may have short- and long-term adverse health effects. Volatile organic compounds are emitted by a wide array of products numbering in the thousands including paints and lacquers, paint strippers, cleaning supplies, and pesticides.

[17]Flaring involves the burning of gas either for safety reasons or because operators do not have the infrastructure to bring the gas to market. For more information on flaring, see GAO, *Federal Oil and Gas Leases: Opportunities Exist to Capture Vented and Flared Natural Gas, Which Would Increase Royalty Payments and Reduce Greenhouse Gases*, GAO-11-34 (Washington, D.C.: Oct. 29, 2010).

Figure 6: Potential Sources and Types of Air Emissions from Oil and Gas Development

Stationary sources

Well

Description: An oil or gas well may be an emissions source, particularly after hydraulic fracturing as fluids and gases flow back to the surface.

Potential emissions: Methane, volatile organic compounds, hydrogen sulfide, and hazardous air pollutants

Storage vessels

Description: Storage vessels are used to hold a variety of liquids, including crude oil, condensate, and produced water.

Potential emissions: Methane, volatile organic compounds, and hazardous air pollutants

Pneumatic controllers

Description: Automated instruments are used for maintaining conditions, such as liquid level, pressure, and temperature.

Potential emissions: Methane, volatile organic compounds, and hazardous air pollutants

Heater treater

Description: Heater treaters are used to remove water from oil.

Potential emissions: Nitrogen oxides and carbon monoxide

Dehydrator

Description: Dehydration removes water that is entrained in the natural gas stream.

Potential emissions: Methane, volatile organic compounds, and hazardous air pollutants

Pits

Description: Pits are used to store freshwater and/or produced water.

Potential emissions: Volatile organic compounds, methane, and hazardous air pollutants

Flaring

Description: Flaring may be used for safety reasons or, in areas where the primary purpose of drilling is to produce oil, gas may be flared because no local market exists for the gas (pictured here).

Potential emissions: Nitrogen oxides, particulate matter, carbon monoxide, carbon dioxide, and sulfur dioxide

Mobile sources

On-road diesel engines

Description: Trucks transport materials to and from the well site.

Potential emissions: Nitrogen oxides, particulate matter, carbon monoxide, and hazardous air pollutants

Off-road diesel engines

Description: Compressors and engines are used to drill and hydraulically fracture the well.

Potential emissions: Nitrogen oxides, particulate matter, carbon monoxide, and hazardous air pollutants

Sources: GAO; photos from GAO except the pneumatic controllers and dehydrator are from BLM, and the pit is courtesy of www.marcellus-shale.us.

Federal Environmental and Public Health Laws Apply to Unconventional Oil and Gas Development but with Key Exemptions

Requirements from eight federal laws apply to the development of oil and gas from unconventional sources. In large part, the same requirements apply to conventional and unconventional oil and gas development. There are exemptions or limitations in regulatory coverage for preventive programs authorized by six of these laws, though EPA generally retains its authorities under federal environmental and public health laws to respond to environmental contamination. States may have regulatory programs related to some of these exemptions or limitations in federal regulatory coverage; state requirements are discussed later in this report.

Eight Federal Environmental and Public Health Laws Apply to Unconventional Oil and Gas Development

Parts of the following eight federal environmental and public health laws apply to unconventional oil and gas development:

- Safe Drinking Water Act (SDWA)

- Clean Water Act (CWA)

- Clean Air Act (CAA)

- Resource Conservation and Recovery Act (RCRA)

- Comprehensive Environmental Response, Compensation, and Liability Act (CERCLA)

- Emergency Planning and Community Right-to-Know Act (EPCRA)

- Toxic Substances Control Act (TSCA)

- Federal Insecticide, Fungicide, and Rodenticide Act (FIFRA)

There are exemptions or limitations in regulatory coverage related to the first six laws listed above. In large part, the same requirements apply to conventional and unconventional oil and gas development. This section discusses each of these laws in brief;[18] for more details about seven of these laws, please see appendixes II through VIII.

---

[18]The National Environmental Policy Act may also apply and is discussed later in this report.

| Safe Drinking Water Act | SDWA is the main federal law that ensures the quality of drinking water.[19] Two key aspects of SDWA that are part of the regulatory framework governing unconventional oil and gas development are the Underground Injection Control (UIC) program and the imminent and substantial endangerment provision. |

UIC Program

Under SDWA, EPA regulates the injection of fluids underground through its UIC program, including the injection of produced water from oil and gas development. The UIC program protects underground sources of drinking water by setting and enforcing standards for siting, constructing, and operating injection wells. Injection wells in the UIC program fall into six different categories based on the types of fluids being injected. The wells used to manage fluids associated with oil and gas production, including produced water, are Class II wells.[20]

EPA officials estimate there are approximately 151,000 permitted Class II UIC wells in operation in the United States. Two types of wells account for nearly all the Class II UIC wells in the United States (see fig. 7), as follows:

- Enhanced recovery wells inject produced water or other fluids or gases into oil- or gas-producing formations to increase the pressure in the formation and force additional oil or gas out of nearby producing wells. EPA documents estimate that about 80 percent of Class II wells are enhanced recovery wells.

- Disposal wells inject produced water or other fluids associated with oil and gas production into formations that are intended to hold the fluids permanently. EPA documents estimate that about 20 percent of Class II wells are disposal wells.[21]

---

[19]Pub. L. No. 93-523 (1974), codified as amended at 42 U.S.C. §§ 300f–300j-26 (2010).

[20]Other classes of UIC wells are used by other industries. For example, Class I wells are for the injection of hazardous, radioactive, and industrial wastes. Class III wells are used for the injection of fluids as part of mining operations, such as for mining salts or uranium.

[21]A third type of Class II UIC well is a hydrocarbon storage well, which injects liquid hydrocarbons into underground formations, such as salt caverns, which can store the hydrocarbons for later use. EPA estimates there are over 100 hydrocarbon storage wells in use in the United States.

Figure 7: Enhanced Recovery and Disposal Wells

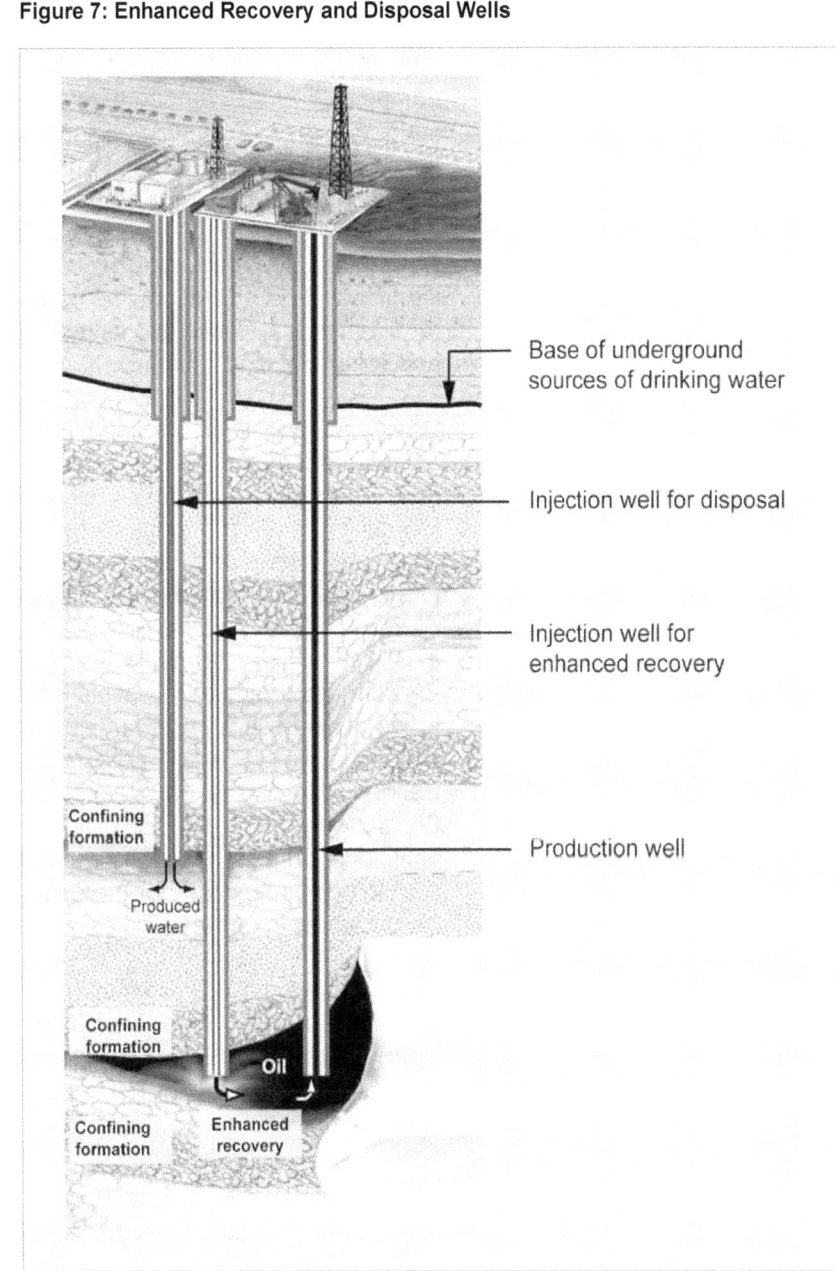

Base of underground sources of drinking water

Injection well for disposal

Injection well for enhanced recovery

Production well

Confining formation

Produced water

Confining formation

Oil

Confining formation

Enhanced recovery

Source: EPA.

GAO-12-874 Unconventional Oil and Gas Development

UIC regulations include minimum federal requirements for most Class II UIC wells; these requirements are generally applicable only where EPA implements the program.[22] For example, for most new Class II UIC wells, an operator[23] must, among other things (1) obtain a permit from EPA or a state, (2) demonstrate that casing and cementing are adequate, and (3) pass an integrity test prior to beginning operation and at least once every 5 years. In addition, when proposing a new Class II UIC well, an operator must identify any existing water or abandoned production or injection wells generally within one-quarter mile of the proposed well. During the life of the Class II UIC well, the operator has to comply with monitoring requirements, including tracking the injection pressure, rate of injection, and volume of fluid injected.

SDWA authorizes EPA to approve by rule a state to be the primary enforcement responsibility—called primacy—for the UIC program, which means that a state assumes responsibility for implementing its program, including permitting and monitoring UIC wells. Generally, to be approved for primacy, state programs must be at least as stringent as the federal program for each of the well classes for which primacy is sought; however, SDWA also includes alternative provisions for primacy related to Class II wells whereby, in lieu of adopting requirements consistent with EPA's Class II regulations, a state can demonstrate to EPA that its program is effective in preventing endangerment of underground sources of drinking water. Five of the six states in our review (Colorado, North Dakota, Ohio, Texas, and Wyoming) have been granted primacy for Class II wells under the alternative provisions. Pennsylvania has not applied for primacy, so EPA directly implements the program there. Please see appendix IX for more information about UIC requirements in the six states in our review.

As discussed, the UIC program regulates the injection of fluids underground. Historically, the UIC program was not used to regulate hydraulic fracturing, even though fracturing entails the injection of fluid underground. In 1994, in light of concerns that hydraulic fracturing of coalbed methane wells threatened drinking water, citizens petitioned EPA

---

[22]See discussion below. According to EPA, some states' UIC requirements for Class II wells are very similar or identical to EPA requirements.

[23]For simplicity, throughout this report we refer to requirements on well operators. In some cases requirements may also apply to well owners.

to withdraw its approval of Alabama's Class II UIC program because the state failed to regulate hydraulic fracturing. The case ended up before the United States Court of Appeals for the Eleventh Circuit, which held that the definition of underground injection included hydraulic fracturing. The court's decision was made in the context of hydraulic fracturing of a coalbed methane formation in Alabama but raised questions about whether hydraulic fracturing would be included in UIC programs nationwide.[24]

In 2005, the Energy Policy Act amended SDWA to specifically exempt hydraulic fracturing from the UIC program, except if diesel fuel is injected as part of hydraulic fracturing. Thus, SDWA as amended continues to authorize regulation of hydraulic fracturing using diesel fuel.[25] EPA officials told us that they do not have data about how frequently companies currently use diesel in hydraulic fracturing.[26] According to EPA officials, EPA recently identified wells for which publicly available data suggest diesel was used in hydraulic fracturing.[27] Since 2005, however, EPA officials said that the agency has not received any permit applications or issued any permits authorizing diesel to be used in hydraulic fracturing. EPA officials also said that they were not aware of any state UIC programs that had issued such permits. EPA headquarters officials said that EPA requires operators conducting hydraulic fracturing operations with diesel fuel to apply for a Class II UIC permit. In May 2012, EPA published draft guidance on how its UIC permit writers should address hydraulic fracturing with diesel in the context of the Class II UIC

---

[24]The court ordered EPA to reconsider its approval of Alabama's program. Legal Environmental Assistance Foundation v. EPA, 118 F.3d 1467, 1478 (11th Cir.1997).

[25]UIC regulations at the time and now provide that "[a]ny underground injection, except into a well authorized by rule or except as authorized by permit issued under the UIC program, is prohibited." 40 C.F.R. 144.11 (2005) (2011). The Energy Policy Act provision did not exempt injections of diesel fuel during hydraulic fracturing from the definition of underground injection. EPA's position is that underground injection of diesel fuel as part of hydraulic fracturing requires a UIC permit or authorization by rule.

[26]In 2003, EPA entered into a memorandum of agreement with three major fracturing service companies in which the companies voluntarily agreed to eliminate diesel fuel in hydraulic fracturing fluids injected into underground sources of drinking water during hydraulic fracturing of coalbed methane wells.

[27]According to EPA officials, EPA Region 3 has sent letters to the relevant operators who may have used diesel in hydraulic fracturing stating they must apply for a Class II UIC permit if they continue using diesel in hydraulic fracturing.

program. The guidance is directed at EPA permit writers in states where EPA directly implements the program; the guidance does not address state-run UIC programs (including five of the six states in our review). EPA's draft guidance is applicable to any oil and gas wells using diesel in hydraulic fracturing (not just coalbed methane wells). The draft guidance provides recommendations related to permit applications, area of review (for other nearby wells), well construction, permit duration, and well closure.

Imminent and Substantial Endangerment Authority

SDWA also gives EPA authority to issue orders when the agency receives information about present or likely contamination of a public water system or an underground source of drinking water that may present an imminent and substantial endangerment to human health. In December 2010, EPA used this authority to issue an emergency administrative order to an operator in Texas alleging that the company's oil and gas production facilities near Fort Worth, Texas, caused or contributed to methane contamination in two nearby private drinking water wells. EPA contended that this methane contamination posed an explosion hazard and therefore was an imminent and substantial threat to human health. EPA's order required the operator to take six actions, specifically: (1) notify EPA whether it intended to comply with the order, (2) provide replacement water supplies to landowners, (3) install meters to monitor for the risk of explosion at the affected homes, (4) conduct a survey of any additional private water wells within 3,000 feet of the oil and gas production facilities, (5) develop a plan to conduct soil and indoor air monitoring at the affected dwellings, and (6) develop a plan to investigate how methane flowed into the aquifer and private drinking water wells. The operator disputed the validity of EPA's order and noted that the order does not provide any way for the company to challenge EPA's findings. Nevertheless, the operator implemented the first three actions EPA listed in the order. In January 2011, EPA sued the operator in federal district court, seeking to enforce the remaining three provisions of the order. In March 2011, the regulatory agency that oversees oil and gas development in Texas held a hearing examining the operator's possible role in the contamination of the water wells and issued an opinion in which it concluded that the operator had not caused the contamination. In March 2012, EPA withdrew the original emergency administrative order, and the operator agreed to continue monitoring 20 private water wells near its production sites for 1 year. According to EPA officials, resolving the lawsuit allows the agency to shift its focus away from litigation and toward a joint EPA-operator effort in monitoring.

For more details about SDWA, please see appendix II.

Clean Water Act

To restore and maintain the nation's waters, CWA authorizes EPA to, among other things, regulate pollutant discharges and respond to spills affecting rivers and streams.[28] Several aspects of CWA are applicable to oil and gas well pad sites, but statutory exemptions limit EPA's regulatory authority. Several elements of CWA and implementing regulations are relevant to oil and gas development from onshore unconventional sources. First, the National Pollutant Discharge Elimination System (NPDES) program regulates industrial sites' wastewater and stormwater discharges to waters of the United States (surface waters).[29] Second, spill reporting and spill prevention and response planning requirements pertain to certain threats to U.S. navigable waters and adjoining shorelines.[30] In addition, under certain circumstances, EPA has response authorities; for example, it can generally bring suit or take other actions to protect the public health and welfare from actual or threatened discharges of oil or hazardous substances to U.S. navigable waters and adjoining shorelines.

NPDES

EPA's NPDES program limits the types and amounts of pollutants that industrial sites, industrial wastewater treatment facilities, and municipal wastewater treatment facilities (often called publicly-owned treatment works or POTWs) can discharge into the nation's surface waters by requiring these facilities to have and comply with permits listing pollutants and their discharge limits. As required by CWA, EPA develops effluent limitations for certain industrial categories based on available control technologies and other factors to prevent or treat the discharge. EPA established multiple

---

[28]The Federal Water Pollution Control Act Amendments of 1972, Pub. L. No. 92-500, § 2, 86 Stat. 816, codified as amended at 33 U.S.C. §§ 1251-1387 (2011) (commonly referred to as the Clean Water Act).

[29]For the purpose of this document, when we use the term "surface waters" in relation to federal regulation, we refer to waters of the United States, including jurisdictional rivers, streams, wetlands, and other waters. State definitions of the term "surface waters" may differ. EPA officials noted that some surface waters may not be jurisdictional for certain CWA provisions.

[30]The scope of jurisdiction for the section 311 oil spill program is broader than that for the NPDES program. The section 311 oil spill program and the NPDES program both have jurisdiction over navigable waters of the United States; Section 311 also provides jurisdiction over spills of oil or hazardous substances into or on adjoining shorelines or that may affect natural resources of the United States, among others.

subcategories for the oil and gas industry; relevant here are: (1) onshore, (2) agricultural and wildlife water use, and (3) stripper wells—that is, wells that produce relatively small amounts of oil.[31]

For the onshore and agricultural and wildlife water use subcategories, EPA established effluent limitations guidelines for direct dischargers that establish minimum requirements to be used by EPA and state NPDES permit writers. Specifically, the onshore subcategory has a zero discharge limit for discharges to surface waters, meaning that no direct discharges to surface waters are allowed. EPA documents explain that this is because there are technologies available—such as underground injection—to dispose of produced water generated at oil and gas well sites without directly discharging them to surface waters. Given that the NPDES permit limit would be "no discharge," EPA officials said that they were unaware of any instances in which operators had applied for these permits. EPA officials did mention, however, an instance in which an operator discharged produced water to a stream and was fined by EPA under provisions in CWA. For example, in 2011, EPA Region 6 assessed an administrative civil penalty against a company managing an oil production facility in Oklahoma for discharging brine and produced water to a nearby stream. The company ultimately agreed to pay a $1,500 fine and conduct an environmental project, which included extensive soil remediation near the facilities.

Effluent limitations guidelines for the agricultural and wildlife water use subcategory cover a geographical subset of wells in the west[32] in which the quality of produced water from the wells is good enough for watering crops and livestock or to support wildlife in streams. The effluent limitations guideline for this subcategory allows such discharges of produced water for these purposes as long as the water meets a minimum quality standard for oil and grease. EPA officials identified 349 facilities with discharge permits in this subcategory. Officials also stated

---

[31]EPA established additional industrial categories in the oil and gas sector for wells in certain near-shore coastal areas, but effluent limitation guidelines for this category are not discussed here, as this report is focused on onshore unconventional oil and gas production.

[32]Specifically, the agricultural and wildlife water use subcategory includes wells located west of the 98th meridian, which extends from approximately the eastern border of North Dakota south through central Texas.

that individual permits may contain limits for pollutants other than oil and grease.

EPA has not established effluent limitations guidelines for stripper wells, and EPA and state NPDES permit writers currently use their best professional judgment to determine the effluent limits for permits on a case-by-case basis. EPA explained in a 1976 *Federal Register* notice that unacceptable economic impacts would occur if the agency developed effluent limitations guidelines for stripper wells and that the agency could revisit this decision at a later date. In July 2012, EPA officials confirmed that the agency currently has no plans to develop an effluent limitations guideline for stripper wells.

EPA also has not established effluent limitations guidelines for coalbed methane wells and EPA and state NPDES permit writers currently use their best professional judgment to determine the effluent limits for permits on a case-by-case basis. EPA officials explained that the process of extracting natural gas from coalbed methane formations is fundamentally different from traditional oil and gas development, partly because of the large volume of water that must be removed from the coalbed methane formation prior to production. Given these differences, coalbed methane wells are not included in any of EPA's current subcategories. EPA announced in 2011 that, based on a multiyear study of the coalbed methane industry, the agency will develop effluent limitations guidelines for produced water discharges from coalbed methane formations. In the course of developing these guidelines, EPA officials told us that they will analyze the economic feasibility of each of the available technologies for disposing of the large volumes of produced water from coalbed methane wells and that EPA plans to issue proposed guidelines in the summer of 2013.

In addition to setting effluent limitations guidelines for direct discharges of pollutants to surface waters, CWA requires EPA to develop regulations that establish pretreatment standards. These standards apply when wastewater is sent to a POTW before being discharged to surface waters, and the standards must prevent the discharge of any pollutant that would interfere with, or pass through, the POTW. To date, EPA has not set pretreatment standards specifically for produced water, though there are some general requirements; for example, discharges to POTWs cannot cause the POTW to violate its NPDES permit or interfere with the treatment process. In October 2011, EPA announced its intention to develop pretreatment standards specific to the produced water from shale gas development. EPA officials told us that the agency intends to conduct a survey and use other methods to collect additional data and information to support this rulemaking. Officials expect to publish the first *Federal Register* notice about the survey by the end of 2012 and to publish a proposed rule in 2014.[33]

In addition to CWA's requirement for NPDES permits for discharges from industrial sites, the 1987 Water Quality Act amended CWA to establish a specific program for regulating stormwater discharges, such as those related to rainstorms, though oil and gas well sites are largely exempt from these requirements. EPA generally requires that facilities get NPDES permits for discharges of stormwater associated with industrial and construction activities, but the Water Quality Act of 1987 specifically exempted oil and gas production sites from permit requirements for stormwater discharges, as long as the stormwater was not contaminated by, for example, raw materials or waste products.[34] As a result of this exemption and EPA's implementing regulations, oil and gas well sites are only required to get NPDES permits for stormwater discharges if the facility has had a discharge of contaminated stormwater that includes a reportable quantity of a pollutant or contributes to the violation of a water quality standard.[35] The 2005 Energy Policy Act expanded the language of the exemption to include construction activities at oil and gas well sites, meaning that uncontaminated stormwater discharges from oil and gas construction sites also do not require NPDES permits. So while other industries must generally obtain NPDES permits for construction activities that disturb an acre or more of land, operators of oil and gas well sites are generally not required to do so.

Spill Reporting and Spill Prevention and Response Planning

CWA prohibits discharges of oil or hazardous substances into U.S. navigable waters or on adjoining shorelines. Specifically, CWA requires facilities—including oil and gas well sites—to report any unpermitted releases of oil or hazardous substances above threshold quantities to the

---

[33]POTWs will be discussed in greater detail later in this report.

[34]The 1987 Water Quality Act also exempted oil and gas processing, treatment, and transmission facilities from permit requirements for stormwater discharges.

[35]EPA has established by regulation threshold amounts of certain pollutants that, if released, trigger reporting requirements; these amounts are known as "reportable quantities." Specifically, the reportable quantities triggering a permit are listed in 40 C.F.R. §§ 117.21, 302.6, 110.6 (2011).

National Response Center, which is managed by the U.S. Coast Guard and serves as the sole federal point of contact for reporting oil and chemical spills in the United States. Oil discharges must be reported if they cause a film or sheen on the surface of the water or shorelines or if they violate water quality standards. The National Response Center shares information about spills with other agencies, including EPA Regional offices, which allows EPA to follow up on reported spills, as appropriate.

CWA also authorized spill prevention and response planning requirements as promulgated in the Spill Prevention, Control, and Countermeasure (SPCC) rule. Facilities that are subject to SPCC rules are required to prepare and implement a plan describing, among other things, how they will control, contain, clean up, and mitigate the effects of any oil discharges that occur. Onshore oil and gas well sites, among others, are subject to this rule if they have total aboveground oil storage capacity greater than 1,320 gallons and could reasonably be expected, based on location, to discharge oil into U.S. navigable waters or on adjoining shorelines.[36] The amount of oil storage capacity at oil and gas well sites tends to vary based on whether the well is being drilled, hydraulically fractured, or has entered production. For example, during drilling at well sites located near these waters, operators generally have to comply with SPCC requirements if fuel tanks for the drilling rig exceed the 1,320 gallon threshold. According to EPA officials, nearly all drill rigs have fuel tanks larger than 1,320 gallons, and so most well sites are subject to the SPCC rule during drilling if they are near these waters. Oil and gas well sites that are subject to the SPCC rule were required to comply by November 2011 or before starting operations.

In accordance with CWA, EPA directly administers the SPCC program rather than delegating authority to states. EPA regulations generally do not require facilities to report SPCC information to EPA, including whether or not they are regulated. As a result, EPA does not know the universe of

---

[36]In addition to having total aboveground oil storage capacity greater than 1,320 gallons, facilities could be required to comply with the SPCC rule if they meet other thresholds, including underground storage capacity of 42,000 gallons, among others.

GAO-12-874 Unconventional Oil and Gas Development

SPCC-regulated facilities.[37] To ensure that regulated facilities are meeting SPCC requirements, EPA Regional personnel may inspect these facilities to evaluate their compliance. EPA officials said that some of these inspections were conducted as follow-up after spills were reported and that most inspections are conducted during the production phase, since drilling and hydraulic fracturing are of much shorter durations, making it difficult for inspectors to visit these sites during those times. According to EPA officials, Regional personnel inspected 120 oil and gas well sites nationwide in fiscal year 2011 and found noncompliance at 105 of these sites. These violations ranged from paperwork inconsistencies to more serious violations, such as a lack of secondary containment around stored oil or failure to implement an SPCC plan (though EPA officials were unable to specifically quantify the number of more serious violations). EPA officials said that EPA has addressed some of the 105 violations through enforcement actions.

Imminent and Substantial Endangerment and Release Response Authorities

CWA also provides EPA with authorities to address the discharge of pollutants and to address actual or threatened discharges of oil or hazardous substances in certain circumstances. For example, under one provision, EPA has the authority to address actual or threatened discharges of oil or hazardous substances into U.S. navigable waters or on adjoining shorelines upon a determination that there may be an imminent and substantial threat to the public health or welfare of the United States, by bringing suit or taking other action, including issuing administrative orders that may be necessary to protect public health and welfare. Under another provision, EPA has authority to obtain records and access to facilities, among other things, in order to determine if a person is violating certain CWA requirements. For example, EPA conducted initial investigations in Bradford County, Pennsylvania, following a 2011 spill of hydraulic fracturing and other fluids that entered a stream. Citing

---

[37]See GAO, *Aboveground Oil Storage Tanks: More Complete Facility Data Could Improve Implementation of EPA's Spill Prevention Program*, GAO-08-482 (Washington, D.C.: Apr. 30, 2008). In that report, we found that EPA has information on only a portion of the facilities subject to the SPCC rule, hindering its ability to identify and effectively target facilities for inspection and enforcement. We recommended that EPA analyze the costs and benefits of the options available to EPA for obtaining key data about the universe of SPCC-regulated facilities, including, among others, a tank registration program similar to those employed by some states. EPA has begun taking action on this recommendation.

its authority under CWA and other laws,[38] EPA requested information from the operator about the incident, including information about the chemicals involved and the environmental effects of the spill. Meanwhile, the Pennsylvania Department of Environmental Protection signed a consent order and agreement with the operator in 2012 that required the operator to pay fines and implement a monitoring plan for the affected stream.

For more details about CWA, please see appendix III.

Clean Air Act

CAA, a federal law that regulates air pollution from mobile and stationary sources, was enacted to improve and protect the quality of the nation's air.[39] Under CAA, EPA sets national ambient air quality standards for the six criteria pollutants—ground level ozone, carbon monoxide, particulate matter, sulfur dioxide, nitrogen oxides, and lead—at levels it determines are necessary to protect public health and welfare. States then develop state implementation plans (SIP) to establish how the state will attain air quality standards, through regulation, permits, policies, and other means. States must obtain EPA approval for SIPs; if a SIP is not acceptable, EPA may assume responsibility for implementing and enforcing CAA in that state. CAA also authorizes EPA to regulate emissions of hazardous air pollutants, such as benzene. In addition, under CAA, EPA requires reporting of greenhouse gas emissions from a variety of sources, including oil and gas wells.

Mobile Sources–Criteria Air Pollutants

In accordance with CAA, EPA has progressively implemented more stringent diesel emissions standards to lower the amount of key pollutants from mobile diesel-powered engines since 1984.[40] These standards apply to a variety of on- and off-road diesel-powered engines, including trucks used in the oil and gas industry to move materials to and from well sites and compressors used to drill and hydraulically fracture wells. Diesel

---

[38]Specifically, EPA cited authorities under CWA section 308, as well as under CERCLA and RCRA.

[39]Clean Air Act Amendments of 1970, Pub. L. No. 91-604, 84 Stat. 1676 (1970), codified as amended at 42 U.S.C. §§ 7401-7671q (2011) (commonly referred to as the Clean Air Act).

[40]See GAO, *Diesel Pollution: Fragmented Federal Programs That Reduce Mobile Source Emissions Could Be Improved*, GAO-12-261 (Washington, D.C.: Feb. 7, 2012).

exhaust contains nitrogen oxides and particulate matter. Emissions standards may set limits on the amount of pollution a vehicle or engine can emit or establish requirements about how the vehicle or engine must be maintained or operated, and generally apply to new vehicles. For example, the most recent emissions standards for construction equipment began to take effect in 2008 and required a 95 percent reduction in nitrogen oxides and a 90 percent reduction in particulate matter from previous standards. EPA estimates that millions of older mobile sources—including on-road and off-road engines and vehicles—remain in use. It is projected that over time, older sources will be taken out of use and be replaced by the lower-emission vehicles, ultimately reducing emissions from mobile sources.

Stationary Sources–Criteria Air Pollutants

New Source Performance Standards (NSPS) apply to new stationary facilities or modifications to stationary facilities that result in increases in air emissions and focus on criteria air pollutants or their precursors.[41] For the oil and gas industry, the key pollutant is volatile organic compounds, a precursor to ground level ozone formation. Prior to 2012, EPA's NSPS were unlikely to affect oil and gas well sites because (1) EPA had not promulgated standards directly targeting well sites[42] and (2) to the extent that EPA promulgated standards for equipment that may be located at well sites, the capacity of equipment located at well sites was generally too low to trigger the requirement. For example, in 1987, EPA issued NSPS for storage vessels containing petroleum liquids; however, the standards apply only to tanks above a certain size, and EPA officials said that most storage tanks at oil and gas sites are below the threshold.

In April 2012, EPA promulgated NSPS for the oil and natural gas production industry which, when fully phased-in by 2015, will require reductions of volatile organic compound emissions at oil and gas well sites, including wells using hydraulic fracturing.[43] Specifically, these new

---

[41]For example, precursors to ground level ozone are nitrogen oxides and volatile organic compounds.

[42]EPA did promulgate standards related to other parts of the oil and gas industry. For example, in 1985, EPA promulgated NSPS that focused on natural gas processing plants, which remove impurities from natural gas to prepare it for use by consumers.

[43]EPA's April 2012 rulemaking also set NSPS for other parts of the oil and natural gas industry, such as for equipment leaks, certain types of compressors, and pneumatic controllers located at natural gas processing plants.

standards are related to pneumatic controllers, well completions, and certain storage vessels as follows:

- *Pneumatic controllers.* According to EPA, when pneumatic controllers are powered by natural gas, they may release natural gas and volatile organic compounds during normal operations. The new standards set limits for the amount of gas (as a surrogate for volatile organic compound emissions) that new and modified pneumatic controllers can release per hour. EPA's regulatory impact analysis for the NSPS estimates that about 13,600 new or modified pneumatic controllers will be required to meet the standard annually; EPA also estimates that the oil and gas production sector currently uses about 400,000 pneumatic controllers.

- *Well completions for hydraulically fractured natural gas wells.* EPA's NSPS for well completions focus on reducing the venting of volatile organic compounds during flowback after hydraulic fracturing. According to EPA's regulatory impact analysis, natural gas well completions involving hydraulic fracturing vent approximately 230 times more natural gas and volatile organic compounds than natural gas well completions that do not involve hydraulic fracturing. The regulatory impact analysis attributes these emissions to the practice of routing flowback of fracture fluids and reservoir gas to a surface impoundment (pit) where natural gas and volatile organic compounds escape to the atmosphere. To reduce the release of volatile organic compounds from hydraulically fractured natural gas wells, EPA's new rule will require operators to use "green completion" techniques to capture and treat flowback emissions so that the captured natural gas can be sold or otherwise used. EPA's regulatory impact analysis for the rule estimates that more than 9,400 wells will be required to meet the new standard annually.[44]

- *Storage vessels.* Storage vessels are used at well sites (and in other parts of the oil and gas industry) to store crude oil, condensate, and produced water. These vessels emit gas and volatile organic compounds when they are being filled or emptied and in association with changes of temperature. EPA's NSPS rule will require storage vessels that emit more than 6 tons per year of volatile organic

---

[44]This estimate includes green completions that are required to occur under the rule, including some that would likely occur voluntarily (e.g., without the rule). EPA estimated that of this total approximately 4,800 such completions would likely occur voluntarily.

compounds to reduce these emissions by at least 95 percent. EPA's regulatory impact analysis for the rule estimates that approximately 300 new storage vessels used by the oil and gas industry will be required to meet the new standards annually. EPA officials said they anticipate that most of these storage vessels will be located at well sites.

Stationary Sources–Hazardous Air Pollutants

EPA also regulates hazardous air pollutants emitted by stationary sources. In accordance with the 1990 amendments to CAA, EPA does this by identifying categories of industrial sources of hazardous air pollutants and requiring those sources to comply with emissions standards, such as by installing controls or changing production practices. These National Emission Standards for Hazardous Air Pollutants (NESHAP) for each industrial source category include standards for major sources, which are defined as sources with the potential to emit 10 tons or more per year of a hazardous air pollutant or 25 tons or more per year of a combination of pollutants, as well as for area sources, which are sources of hazardous air pollutants that are not defined as major sources. Generally, EPA or state regulators can aggregate emissions from related or nearby equipment to determine whether the unit or facility should be regulated as a major source. However, in determining whether the oil or gas well is a major source of hazardous air pollutants, CAA expressly prohibits aggregating emissions from oil and gas wells (with their associated equipment) and emissions from pipeline compressors or pumping stations.

EPA initially promulgated a NESHAP for oil and natural gas production facilities for major sources in 1999 and promulgated amendments in April 2012. NESHAPs generally identify emissions points that may be present at facilities within each industrial source category. The source category for oil and natural gas production facilities includes oil and gas well sites and other oil and gas facilities, such as pipeline gathering stations and natural gas processing plants. The NESHAP for the oil and natural gas production facilities major source category includes emissions points (or sources) that may or may not normally be found at well sites at sizes that would tend to meet the major source threshold. EPA officials in each of the four Regions we contacted were unaware of any specific examples of oil and natural gas wells being regulated as major sources of hazardous air pollutants before the April 2012 amendments. These amendments, however, changed a key definition used to determine whether a facility (such as a well site) is a major source. Specifically, EPA modified the definition of the term "associated equipment" such that emissions from all storage vessels and glycol dehydrators (used to remove water from gas) at a facility will be counted toward determining whether a facility is a major source. EPA's regulatory impact analysis and other technical support documents for the April 2012 amendments did not estimate how many oil and natural gas well sites would be considered major sources under the new definition.

EPA also promulgated a NESHAP for oil and natural gas production facilities for area sources in 2007. The 2007 area source rule addresses emissions from one emissions point, triethylene glycol dehydrators, which are used to remove water from gas. Triethylene glycol dehydrators can be located at oil and gas well sites or other oil and gas facilities, such as natural gas processing plants. Area sources are required to notify EPA that they are subject to the rule, but EPA does not track whether the facilities providing notification are well sites or other oil and natural gas facilities, so it is difficult to determine to what extent oil and gas well sites are subject to the area source NESHAP.[45]

In addition to specific programs for regulating hazardous air pollutants, CAA establishes that operators of stationary sources that produce, process, store, or handle listed or extremely hazardous substances have a general duty to identify hazards that may result from accidental releases, take steps needed to prevent such releases, and minimize the consequences of such releases when they occur. Methane is one of many hazardous substances of concern due to their flammable properties. Some EPA Regional officials said that they use infrared video cameras to conduct inspections to identify leaks of methane from storage tanks or other equipment at well sites. For example, EPA Region 6 officials said they have conducted 45 inspections at well sites from July 2010 to July 2012 and issued 10 administrative orders related to violations of the CAA general duty clause.[46] EPA headquarters officials said that all well sites are required to comply with the general duty clause but that EPA prioritizes and selects sites for inspections based on risk.

---

[45]In addition to NESHAPs specific to the oil and natural gas production industrial source category, EPA promulgated other NESHAPs that could apply to oil and gas well sites depending on the types of equipment in use and their size. See appendix IV for more details.

[46]EPA Region 6 includes the states of Arkansas, Louisiana, New Mexico, Oklahoma, and Texas.

GAO-12-874 Unconventional Oil and Gas Development

CAA also requires EPA to publish regulations and guidance for chemical accident prevention at facilities using substances that pose the greatest risk of harm from accidental releases; the regulatory program is known as the Risk Management Program. The extent to which a facility is subject to the Risk Management Program depends on the regulated substances present at the facility and their quantities, among other things. EPA's list of regulated substances and their thresholds for the Risk Management Program was initially established in 1994 and has been revised several times. The regulated chemicals that may be present at oil and gas well sites include components of natural gas (e.g., butane, propane, methane, and ethane). However, a 1998 regulatory determination from EPA provided an exemption for naturally-occurring hydrocarbon mixtures (i.e., crude oil, natural gas, condensate, and produced water) prior to entry into a natural gas processing plant or petroleum refinery; EPA explained at the time that these chemicals do not warrant regulation and that the general duty clause would apply in certain risky situations.[47] Since naturally-occurring hydrocarbons at well sites generally have not entered a processing facility, they are not included in the threshold determination of whether the well site should be subject to the Risk Management Program. EPA officials said that generally, unless other flammable or toxic regulated substances were brought to the site, well sites would not trip the threshold quantities for the risk management regulations. In September 2011, the U.S. Chemical Safety and Hazard Investigation Board (Chemical Safety Board) released a report describing 26 incidents involving fatalities or injuries related to oil and gas storage tanks located at well sites from 1983 through 2010.[48] The report found that these accidents occurred when the victims—all young adults—gathered at rural unmanned oil and gas storage sites lacking fencing and warning signs and concluded that such sites pose a public safety risk. The report also noted that exploration and production storage tanks are exempt from the Risk Management Program requirements of CAA and recommended that EPA publish a safety alert to owners and operators of exploration and

[47]In addition, a 1999 law provided an exemption for flammable substances being used as fuel; this exemption applies to any type of facility using fuel. Chemical Safety Information, Site Security and Fuels Regulatory Relief Act, Pub. L. No. 106–40 § 2, 113 Stat 207 (1999).

[48]The Chemical Safety Board is an independent federal agency investigating chemical accidents to protect workers, the public, and the environment. See U.S. Chemical Safety and Hazard Investigation Board, *Investigative Study Final Report: Public Safety at Oil and Gas Storage Facilities*, Report No. 2011-H-1 (September 2011).

production facilities with flammable storage tanks advising them of their CAA general duty clause responsibilities, and encouraging specific measures to reduce these risks.[49] The Chemical Safety Board requested that EPA provide a response stating how EPA will address the recommendation. EPA responded in June 2012, stating its intent to comply with the recommendation.

Stationary Sources–Greenhouse Gas Reporting

As of 2012, oil and natural gas production facilities are required to report their greenhouse gas emissions to EPA on an annual basis as described in EPA's greenhouse gas reporting rule. According to EPA documents, oil and gas well sites may emit greenhouse gases, including methane and carbon dioxide, from sources including: (1) combustion sources, such as engines used on site, which typically burn natural gas or diesel fuel, and (2) indirect sources, such as equipment leaks and venting.[50] The greenhouse gas reporting rule requires oil and gas production facilities (defined in regulation as all wells in a single basin that are under common ownership or control) that emit more than 25,000 metric tons of carbon dioxide equivalent at the basin level to report their annual emissions of carbon dioxide, methane, and nitrous oxide from equipment leaks and venting, gas flaring, and stationary and portable combustion. EPA documents estimate that emissions from approximately 467,000 onshore wells are covered under the rule.

For more details about CAA, please see appendix IV.

Resource Conservation and Recovery Act

RCRA, passed in 1976, established EPA's authority to regulate the generation, transportation, treatment, storage, and disposal of hazardous wastes.[51] Subsequently, the Solid Waste Disposal Act Amendments of 1980 created a separate process by which oil and gas exploration and

---

[49]The Chemical Safety Board also noted that exploration and production storage tanks are exempt from the security requirements of CWA's spill prevention program.

[50]Other major greenhouse gases covered by EPA's greenhouse gas reporting rule include hydrofluorocarbons, perfluorocarbons, and sulfur hexafluoride.

[51]Pub. L. No. 94-580, 90 Stat. 2795 (1976) (amending the Solid Waste Disposal Act, but generally referred to as RCRA), codified as amended at 42 U.S.C. §§ 6901-6992k (2011). RCRA also created a framework in which states are largely responsible for solid (i.e., nonhazardous) waste regulations, including treatment and land disposal of these wastes. State solid waste provisions will be discussed in greater detail later in this report.

production wastes, including those originating within a well, would not be regulated as hazardous unless EPA conducted a study of wastes associated with oil and gas development and then determined that such wastes warranted regulation as hazardous waste, followed by congressional approval of the regulations.[52] EPA conducted the study and, in 1988, issued a determination that it was not warranted to regulate oil and gas exploration and production wastes as hazardous. Based on this EPA determination, drilling fluids, produced water, and other wastes associated with the exploration, development, or production of oil or gas are not regulated as hazardous. According to EPA guidance issued in 2002, these exempt wastes include wastes that come from within the well, as well as wastes generated from field operations.[53] Conversely, wastes generated from other activities at well sites may be regulated as hazardous. For example, discarded unused hydraulic fracturing fluids and painting wastes, among others, may be present at well sites and are "non-exempt," and could be regulated as hazardous, depending on the specific characteristics of the wastes. Facilities that generate more than 100 kilograms (220 pounds) of hazardous waste per month are regulated as generators of hazardous waste and, among other things, are required to have an EPA identification number and to use the RCRA manifest system for tracking hazardous waste.[54] Facilities generating smaller quantities of hazardous waste are not subject to these requirements.[55] EPA headquarters officials said they do not have data on how many well sites may be hazardous waste generators, but that states may have more information about quantities of hazardous wastes at well sites. As such,

---

[52]Specifically, oil and gas exploration and production waste includes drilling fluids, produced waters, and other wastes associated with the exploration, development, or production of crude oil or natural gas, among other things.

[53]Field operations include, for example, water separation, demulsifying, degassing, and storage at well sites.

[54]Small quantity generators are those generating between 100 and 1,000 kilograms of hazardous waste per month, and large quantity generators are those generating more than 1,000 kilograms per month. Both small and large quantity generators are required to obtain an EPA identification number and are subject to certain regulations. Facilities generating less than 100 kilograms of hazardous waste per month are considered conditionally exempt small quantity generators provided they meet certain conditions, and if so, do not need to obtain an identification number.

[55]These conditionally exempt small quantity generators are subject to limited generator waste management standards, namely to identify their hazardous waste, comply with storage limit requirements, and ensure waste treatment or disposal in a proper facility.

we asked state officials responsible for waste programs whether they were aware of well sites being classified as small-quantity hazardous waste generators, and officials in all six states we reviewed indicated that they were unaware of well sites having sufficient quantities of hazardous wastes to be subject to those regulations.

In September 2010, the Natural Resources Defense Council submitted a petition to EPA requesting that the agency regulate waste associated with oil and gas exploration and production as hazardous. The petition asserts that EPA should revisit the 1988 determination not to regulate these wastes as hazardous because, among other things, EPA's underlying assumptions concerning the availability of alternative disposal practices, the adequacy of state regulations, and the potential for economic harm to the oil industry are no longer valid. According to EPA officials, the agency is currently reviewing the information provided in the petition but does not have a time frame for responding.

RCRA also authorizes EPA to issue administrative orders, among other things, in cases where handling, treatment, or storage of hazardous or solid waste may present an imminent and substantial endangerment to health or the environment. EPA has used RCRA's imminent and substantial endangerment authorities related to oil and gas well sites. For example, EPA Region 8 issued RCRA imminent and substantial endangerment orders to operators in Wyoming after discovering that pits near oil production sites were covered with oil and posed a hazard to birds.

For more details about RCRA, please see appendix V.

Comprehensive Environmental Response, Compensation, and Liability Act

Congress passed CERCLA in 1980 to protect public health and the environment by addressing the cleanup of hazardous substance releases.[56] CERCLA establishes a system governing the reporting and cleanup of releases of hazardous substances and provides the federal government the authority to respond to actual and threatened releases of hazardous substances, pollutants, and contaminants that may endanger public health and the environment. CERCLA requires operators of oil and gas sites to report certain releases of hazardous substances and gives

---

[56]Pub. L. No. 96-510, 94 Stat. 2767 (1980), codified as amended at 42 U.S.C. §§ 9601-9675 (2012).

EPA authority to respond to certain releases but excludes releases of petroleum (e.g., crude oil and other petroleum products) from these provisions. As previously discussed, releases of petroleum products are covered by CWA if the release threatens U.S. navigable waters or adjoining shorelines. EPA officials identified some instances of petroleum spills in dry areas that did not reach surface waters and explained that EPA had no role related to the investigation or cleanup of these incidents. We identified regulatory provisions in five of six states requiring cleanup of oil spills even if they do not reach surface waters.

For hazardous substances, CERCLA has two key elements relevant for the unconventional oil and gas industry: release reporting and EPA's investigative and response authority. Similar to the requirements to report oil spills under CWA, CERCLA requires operators to report releases of hazardous substances above reportable quantities to the National Response Center. The National Response Center shares information about spills with other agencies, including EPA Regional offices, which allows EPA the opportunity to follow up on reported spills. EPA also has investigative and response authority under CERCLA, including provisions allowing EPA broad access to information and the authority to enter property to conduct an investigation or a removal of contaminated material. EPA has the following authorities, among others:

- *Investigative.* EPA may conduct investigations—including activities such as monitoring, surveying, and testing—in response to actual or threatened releases of hazardous substances or pollutants or contaminants. EPA can also require persons to provide information about alleged releases or threat of release. EPA officials described several instances in which the agency used CERCLA's investigative and information gathering authorities relating to alleged hazardous substance releases from oil and gas well sites. For example, EPA used CERCLA authority to investigate private water well contamination potentially related to nearby shale gas well sites in Dimock, Pennsylvania. In addition, EPA is currently using the same authority to investigate private water well contamination potentially related to tight sandstone well sites in Pavillion, Wyoming.

- *Response.* EPA has the authority to respond to releases of hazardous substances itself and to issue administrative orders requiring a company potentially responsible for a release of hazardous substances, which may pose an imminent and substantial endangerment, to take response actions, as well as to seek relief in a federal court. EPA officials could not provide a recent example where the agency used this authority to issue an administrative order at a well site, but EPA used the response authority to conduct sampling and to provide temporary drinking water to residents with contaminated wells in Dimock, Pennsylvania.

For more details about CERCLA, please see appendix VI.

Emergency Planning and Community Right-to-Know Act

Among other things, EPCRA provides individuals and their communities with access to information regarding storage or release of certain chemicals in their communities.[57] Two provisions of EPCRA—release notification and chemical storage reporting—apply to oil and gas well sites. The release notification provisions require companies that produce, use, or store certain chemicals to notify state and local emergency planning authorities of certain releases that would affect the community.[58] Spills that are strictly on-site would not have to be reported under EPCRA but may still have to be reported to the National Response Center under provisions of CWA or CERCLA. In addition, companies would have to comply with EPCRA's chemical storage reporting provisions, which require facilities storing or using hazardous or extremely hazardous chemicals over certain thresholds to submit an annual inventory report including detailed chemical information to state and local emergency planning authorities and the local fire department.[59] When asked whether oil and gas well sites would commonly trigger EPCRA's release notification and chemical storage reporting requirements, EPA officials said these requirements could be triggered at every well site.

EPCRA also established the Toxics Release Inventory (TRI)—a publicly available database containing information about chemical releases from more than 20,000 industrial facilities—but EPA regulations for the TRI do

---

[57]Pub. L. No. 99-499, 100 Stat. 1728 (1986), codified at 42 U.S.C. §§ 11001–11050 (2012).

[58]Three types of releases must be reported: (1) release of extremely hazardous substances for which notification is also required under CERCLA § 103(c), (2) release of extremely hazardous substances for which notification is not required under CERCLA § 103(c), but above reporting thresholds and subject to additional conditions, and (3) release of other hazardous substances for which notification is also required under CERCLA § 103(c), subject to CERCLA reporting thresholds or 1 pound default threshold. EPCRA § 304(a), 42 U.S.C. §§ 11004(a) (2011).

[59]Specifically, the thresholds are (1) more than 500 pounds or the threshold planning quantity, whichever is lower, of extremely hazardous substances, or (2) more than 10,000 pounds of other hazardous chemicals.

not require oil and gas well sites to report to TRI. Specifically, these provisions of EPCRA generally require certain facilities that manufacture, process, or otherwise use any of more than 600 listed chemicals to report annually to EPA and their respective states on chemicals used above threshold quantities; the amounts released to the environment; and whether they were released into the air, water, or soil. EPCRA specified certain industries subject to the reporting requirement—which did not include oil and gas exploration and development—and also provided authority for EPA to add or delete industries going forward.[60] EPA issued regulations to implement the TRI in 1988 and chose not to change the list of industries subject to the provision at that time. In 1997, EPA promulgated a rule adding seven industry groups to the list of industries required to report releases to the TRI, including coal mining and electrical utilities that combust coal and/or oil.[61] In developing the 1997 rule, EPA considered including oil and gas exploration and production but did not do so because, according to EPA's notice in the *Federal Register* for the final rule, there were concerns about how "facility" would be defined for this industry. At that time, EPA's stated rationale was that the oil and gas exploration and production industry is unique in that it may have related activities over a large geographic area and, while together these activities may involved the management of chemicals regulated by the TRI program, taken at the smallest unit—an individual well—the chemical and other thresholds are unlikely to be met.[62] According to EPA officials, EPA is in the preproposal stage of developing a new rule to add additional industrial sectors into the TRI program but is not planning to include the oil and gas exploration and production industry.[63] EPA officials said that adding oil and gas well sites would likely provide an incomplete picture of

[60]In addition to identifying industries, EPCRA specifies that reporting requirements apply to owners or operators of covered facilities: (1) with 10 or more full-time employees and (2) that manufactured, processed, or otherwise used a listed toxic chemical in excess of the reporting threshold during the calendar year.

[61]The complete list of industries added by EPA in 1997 includes metal mining, coal mining, electrical utilities that combust coal and/or oil for the purpose of generating power for distribution in commerce, refuse processing or destruction facilities regulated under RCRA's hazardous waste provisions, chemical wholesalers, petroleum terminals, and bulk stations and solvent recovery services.

[62]Other thresholds include the number of employees at the facility.

[63]Officials said that EPA is also considering steam generation from coal and/or oil, petroleum bulk storage, iron ore mining, phosphate mining, large dry cleaning, and solid waste combustors and incinerators.

the chemical uses and releases at these sites and would, therefore, be of limited utility in providing information to communities.

For more details about EPCRA, please see appendix VII.

Toxic Substances Control Act

TSCA authorizes EPA to regulate the manufacture, processing, use, distribution in commerce, and disposal of chemical substances and mixtures.[64] TSCA provides EPA with several authorities by which EPA may assess and manage chemical risks, including the authority to (1) collect information about chemical substances, (2) require companies to conduct testing on chemical substances, and (3) take action to protect adequately against unreasonable risks.[65] TSCA allows chemical companies to assert confidentiality claims on information provided to EPA; if the information provided meets certain criteria, EPA must protect it from disclosure to the public.

EPA maintains a list of chemicals that are or have been manufactured or processed in the United States, called the TSCA inventory. Of the over 84,000 chemicals currently in the TSCA inventory, about 62,000 were already in commerce when EPA began reviewing chemicals in 1979. Since then, EPA has reviewed more than 45,000 new chemicals, of which approximately 20,000 were added to the inventory after chemical companies began manufacturing them. As part of EPA's *Study on the Potential Impacts of Hydraulic Fracturing on Drinking Water Resources*, EPA is currently analyzing information provided by nine hydraulic fracturing service companies, including a list of chemicals used in hydraulic fracturing operations. EPA officials said that they expect most of

---

[64]Pub. L. No. 94-469, 90 Stat. 2003 (1976), codified as amended at 15 U.S.C. §§ 2601-2692 (2012). TSCA addresses those chemicals manufactured or imported into the United States, but it generally excludes certain substances, such as pesticides, which are regulated under the Federal Insecticide, Fungicide, and Rodenticide Act, and any food, food additive, drug, cosmetic, or device regulated under the Federal Food, Drug, and Cosmetics Act. Hereinafter, references to chemical substances in discussions of TSCA are intended to include mixtures.

[65]These authorities are conditional on EPA making certain findings. For example, prior to requiring testing under section 4, TSCA requires EPA to make findings (1) regarding the risk of injury to health or the environment or regarding human exposure, (2) that existing data are insufficient, and (3) that testing with respect to such effects is necessary to develop needed data. Regarding actions to protect adequately against unreasonable risks, examples of EPA actions include prohibiting or limiting the manufacture, processing, or distribution in commerce of chemical substances or by placing restrictions on chemical uses.

---

the chemicals disclosed by the service companies to appear on the TSCA inventory list, provided that chemicals are not classified solely as pesticides. EPA officials do not expect to be able to compare the list of chemicals provided by the nine hydraulic fracturing service companies to the TSCA inventory until the release of a draft report of the *Study on the Potential Impacts of Hydraulic Fracturing on Drinking Water Resources* for peer review, expected in late 2014.

In August 2011, EPA received a petition from the environmental group Earthjustice and 114 others asking the agency to exercise TSCA authorities and issue rules to require manufacturers and processors of chemicals used in oil and gas exploration and production to develop and provide certain information to EPA.[66] According to the petition, EPA and the public currently lack adequate information about the health and environmental effects of chemicals used in oil and gas exploration and production, and EPA should exercise its TSCA authorities to ensure that chemicals used in oil and gas exploration and production do not present an unreasonable risk of harm to health and the environment. In a letter to the petitioners, EPA granted the petition in part, stating there is value in beginning a rulemaking process under TSCA to obtain data on chemical substances used in hydraulic fracturing. EPA's letter also stated that the TSCA proposal would focus on providing an aggregate picture of the chemical substances used in hydraulic fracturing, which would complement and not duplicate well-by-well disclosure programs that exist in some states. The letter also indicates that the agency is drafting an Advance Notice of Proposed Rulemaking on this issue. As of August 31, 2012, EPA has not released a publication date for this proposed rulemaking. EPA also intends to convene a stakeholder process to gather additional information for use in developing a proposed rule.

For more details about TSCA, please see appendix VIII.

Federal Insecticide, Fungicide, and Rodenticide Act

FIFRA, as amended, mandates that EPA administer pesticide registration requirements and authorizes EPA to regulate the use, sale, and distribution of pesticides to protect human health and preserve the

---

[66]Earthjustice et al., Letter to Lisa P. Jackson, EPA Administrator, re: Citizen Petition under Toxic Substances Control Act Regarding the Chemical Substances and Mixtures Used in Oil and Gas Exploration or Production, Aug. 4, 2011.

GAO-12-874 Unconventional Oil and Gas Development

environment.[67] FIFRA requires that EPA register new pesticides; pesticide registration is a very specific process that is not valid for all uses of a particular chemical. Instead, each registration describes the chemical and its intended use (i.e., the crops/sites on which it may be applied), and each use must be supported by research data. According to EPA officials, some pesticides registered under FIFRA are used in hydraulic fracturing, and EPA has approved registrations of some pesticides for this purpose. According to a report about shale gas development by the Ground Water Protection Council,[68] operators may use pesticides to kill bacteria or other organisms that may interfere with the hydraulic fracturing process. For example, glutaraldehyde may be used by operators to eliminate bacteria that produce byproducts that cause corrosion inside the well and was reregistered for this purpose by EPA in 2007.[69]

Exemptions Are Related to Preventive Programs

As discussed above, in six of the eight federal environmental and public health laws identified, there are exemptions or limitations in regulatory coverage related to the oil and gas exploration and production industry (there are two exemptions related to CAA). All of these exemptions are related to programs designed to prevent pollution (see table 2). For example, under CWA, EPA generally requires permits for stormwater discharges at construction sites, which prevents sediment from entering nearby streams. However, the Water Quality Act of 1987 and Energy Policy Act of 2005 largely exempted the oil and gas exploration and production sector from these stormwater permitting requirements. Four of the exemptions are statutory (related to SDWA, CWA, CAA, and CERCLA), and three are related to regulatory decisions made by EPA (related to CAA, RCRA, and EPCRA). States may have regulatory programs related to some of these exemptions or limitations in federal regulatory coverage. For example, although oil and gas exploration and production wastes are not regulated under RCRA as hazardous, which reduces the federal role in management of such wastes, they are

---

[67]The Federal Environmental Pesticide Control Act, Pub. L. No. 92-516, 86 Stat. 973 (1972) (amending FIFRA), codified as amended at 7 U.S.C. §§ 136-136y (2012).

[68]Ground Water Protection Council and ALL Consulting. "Modern Shale Gas Development in the United States: A Primer." Prepared for the Department of Energy and National Energy Technology Laboratory. April 2009.

[69]Glutaraldehyde is also used as a disinfectant for medical and dental equipment, in water treatment systems, and as a preservative.

nonetheless solid wastes. State regulations may govern management of solid waste, and certain EPA regulations address minimum requirements for how solid waste disposal facilities should be designed and operated.

Table 2: Exemptions or Limitations in Regulatory Coverage for the Oil and Gas Exploration and Production Industry in Six Environmental Laws

Law	Description of exemption or limitation in regulatory coverage	Source	Type of program related to exemption or limitation in regulatory coverage	
			Preventive	Response
SDWA	Hydraulic fracturing with fluids other than diesel fuel does not require a UIC permit.	Statutory (2005)	X	
CWA	Federal stormwater permits are not required for uncontaminated stormwater at oil and gas construction sites or at oil and gas well sites.	Statutory (1987, 2005)	X	
CAA	Emissions of hazardous air pollutants from oil and gas wells and their associated equipment may not be aggregated together or with those of pipeline compressors or pump stations to determine whether they are a major source.	Statutory (1990)	X	
	In the Risk Management Program, many naturally-occurring hydrocarbons in oil and gas are not included in the threshold determination of whether a facility should be regulated.	Regulatory/EPA decision (1988)	X	
RCRA	Oil and gas exploration and production wastes are not regulated as hazardous waste.	Regulatory/EPA decision (1988)	X	
CERCLA	Liability and reporting provisions do not apply to injections of fluids authorized by state law for production, enhanced recovery, or produced water.	Statutory (1980)	X	
EPCRA	Oil and gas well operations are not required to report releases of listed chemicals to the TRI.	Regulatory/EPA decision (1997)	X	

Source: GAO.

Note: In some cases, states may have requirements in these areas. State requirements are discussed in the next section of this report.

The exemptions do not limit the authorities EPA has under federal environmental and public health laws to respond to environmental contamination. Table 3 lists EPA authorities that may be applicable when conditions or events at a well site present particular risk to the environment or human health. As noted throughout this report, EPA has used several of these authorities at oil and gas wells. For example, as discussed above, EPA Region 8 has used RCRA's imminent and substantial endangerment authorities to issue RCRA imminent and substantial endangerment orders to operators in Wyoming after discovering that pits near oil production sites were covered with oil and posed a hazard to birds. Similarly, as discussed above, EPA is using CERCLA's response authority to investigate private water well contamination in Pavillion, Wyoming.

Whether an authority is available depends on requisite conditions being met in a given instance. EPA officials said that, in some instances, response authorities of multiple federal environmental laws could be used to address a threat to public health or the environment. In 2001, EPA and the Department of Justice developed a memo advocating that officials consider the specifics of a situation and use the most appropriate authority.[70] See appendixes II through VI for a more detailed discussion of these authorities.

---

[70]See EPA, Memorandum, Use of CERCLA § 106 To Address Endangerments that May Also Be Addressed Under Other Environmental Statutes, App. A (2001).

Table 3: Key EPA Response Authorities Relevant to Oil and Gas Well Sites[a]

Law	Key response authorities	Situation to which authority may apply
Imminent and substantial endangerment and general response authorities		
SDWA	Imminent and substantial endangerment (§ 1431)	Contaminant present in or likely to enter a public water system or an underground source of drinking water
CWA	Imminent and substantial endangerment (§ 504)	Source(s) of pollution, including discharge of pollutant to surface waters
	Response authority; imminent and substantial threat (§ 311)	Actual or threatened discharge of oil or hazardous substances into U.S navigable waters or on adjoining shorelines
CAA[b]	Imminent and substantial endangerment (§ 112(r)(9))	Accidental release to the air of regulated substance
RCRA	Imminent and substantial endangerment (§ 7003)	Past or present handling, storage, treatment, transportation, or disposal of any solid waste or hazardous waste
CERCLA	Response authority (§ 104(a))	Actual or threatened release of any hazardous substance
	Imminent and substantial endangerment (§ 104(a))	Actual or threatened release of any hazardous substance or pollutant or contaminant (other than petroleum)
	Imminent and substantial endangerment (§ 106(a))	Actual or threatened release of a hazardous substance from a facility[c]
Access, information, and inspection authorities		
SDWA	Access to records and to inspect facilities (§ 1445(b))	Persons and facilities subject to UIC program requirements
CWA	Access to records and to inspect facilities; require reports (§ 308(a))	Location of effluent source
	Access to records and to inspect facilities; ability to require provision of information (§ 311(m)(2))	Persons and facilities subject to section 311, such as SPCC program requirements
CAA	Access to records; ability to require provision of information (§ 114(a))	Person who owns or operates any emission source, among others
RCRA	Access to records and to inspect facilities (§ 3007)	Persons that have generated, stored, treated, transported, disposed of, or otherwise handled hazardous wastes[d]
CERCLA	Access to records and to inspect facilities; ability to require provision of information (§ 104(e))	Location of actual or threatened release, generation, storage, treatment, or disposal of any hazardous substance or pollutant or contaminant (other than petroleum)
TSCA	Access to inspect facilities and onsite records (§ 11)	In relevant part, facilities where chemical substances are manufactured, processed, stored, or held before or after their distribution or sale

Source: GAO.

[a]The table lists selected EPA authorities that may be applicable when conditions or events at a well site present particular risk to the environment or human health. Whether a particular authority is applicable depends upon the facts of the situation meeting all prerequisite conditions. EPA has other authorities not listed in the table, including the ability to require certain persons to provide information, such as information to aid in developing plans or standards, and the ability to sample emissions or effluent. EPA also has authorities by which it may enforce requirements and address violations of the programs it administers.

[b]In addition, CAA section 303 provides EPA a general imminent and substantial endangerment authority to address emission of air pollutants, where conditions are met.

[c]Generally, CERCLA section 104 authorizes EPA to take various actions to respond to a release, whereas section 106 authorizes EPA to require potentially responsible parties to do so.

[d]EPA interprets this provision of RCRA to include solid waste that EPA reasonably believes may pose a hazard when improperly managed.

States in Our Review Implement Additional Requirements and Recently Updated Some Requirements

The six states in our review implement additional requirements governing a number of activities associated with oil and gas development. One of the states—Pennsylvania—is also part of the Delaware River Basin Commission—a regional commission that implements additional requirements. All six states have updated some aspects of their requirements in recent years.

States in Our Review Implement Additional Requirements and Certain Federal Requirements

In addition to implementing and enforcing certain aspects of federal requirements with EPA approval and oversight, the six states in our review implement additional requirements governing a number of activities associated with oil and gas development. State requirements often do not explicitly differentiate between conventional and unconventional development but, in recent years, states have begun to promulgate some requirements that apply specifically to unconventional development. States have regulatory requirements related to a variety of activities involved in developing unconventional reservoirs, including siting and site preparation; drilling, casing, and cementing; hydraulic fracturing; well plugging; site reclamation; waste management and disposal; and managing air emissions. Table 4 compares selected state requirements and related federal environmental and public health requirements; a more comprehensive table is available in appendix X. Several studies noted that development practices and state requirements may vary based on a number of factors, including geology, climate, and the type of resource being developed.[71] We did not assess whether all requirements are appropriate for all states as part of this review.

---

[71]Charles G. Groat, Ph.D. and Thomas W. Grimshaw, Ph.D., *Fact-Based Regulation for Environmental Protection in Shale Gas Development* (Austin, Texas: The Energy Institute, The University of Texas at Austin, February 2012). Ground Water Protection Council and ALL Consulting. "Modern Shale Gas Development in the United States: A Primer." Prepared for the Department of Energy and National Energy Technology Laboratory. April 2009.

Table 4: Key Federal Environmental and Public Health Requirements and State Requirements for Oil and Gas Production Wells

Area of regulation	EPA environmental and public health requirements	Requirements of six states reviewed
Siting and site preparation		
Identification or testing of water wells prior to drilling of production wells	No	1 of 6 (Wyoming) [identification alone] 2 of 6 (Colorado, Ohio) [identification and testing][a]
Required setbacks from water sources	No	5 of 6 (Colorado, North Dakota, Pennsylvania, Ohio, Wyoming)
Stormwater permitting	Effectively no[b]	4 of 6 (Colorado, North Dakota, Pennsylvania, Wyoming)
Drilling, casing, and cementing		
Requirements relating to cementing/casing plans	No[c]	6 of 6
Prescribed placement of surface casing relative to groundwater zones	No[c]	6 of 6
Hydraulic fracturing		
Requirements to disclose information on fracturing fluids	No[d]	6 of 6
Well plugging		
Requirements for notification, plugging plan or method, witnessing, and reporting	No	6 of 6
Programs to plug wells that are not properly plugged and have been abandoned	No	6 of 6
Site reclamation		
Requirements for backfilling, regrading, recontouring, and alleviating compaction of soil	No	6 of 6
Revegetation requirements	No	5 of 6 (Colorado, North Dakota, Ohio, Pennsylvania, Wyoming)
Waste management		
Pit lining requirements	No	5 of 6 (Colorado, North Dakota, Pennsylvania, Texas, Wyoming)
Options for waste disposal:		
Underground injection	Yes (SDWA)	5 states have their own requirements (Colorado, North Dakota, Ohio, Texas, Wyoming); EPA implements the program in Pennsylvania
Direct discharge to surface water	Yes (CWA – certain discharges prohibited, others subject to conditions and permit)	Surface discharges are allowed in certain cases in 3 western states (Colorado, Texas, and Wyoming)

GAO-12-874 Unconventional Oil and Gas Development

Area of regulation	EPA environmental and public health requirements	Requirements of six states reviewed
Requirements for discharge to POTWs or Centralized Waste Treatment (CWT) facilities	Pretreatment standards for shale gas wastewater under development (CWA)	Disposal at POTWs is an option in two states (Ohio, Pennsylvania)[e] Disposal at CWT facilities is an option in 3 states (Colorado, Pennsylvania, Wyoming)
Recycling or other reuse	Yes (CWA – certain produced water discharges)	6 of 6 states allow recycling or other reuse
Solid waste disposal	Effectively no[f]	Yes
Hazardous waste disposal	Effectively no[g]	No
Managing air emissions		
Requirements for criteria pollutants	Certain CAA provisions apply	5 of 6 states have permitting or registration programs (Colorado, North Dakota, Ohio, Texas, Wyoming)
Requirements for hazardous air pollutants	Certain CAA provisions apply	State permitting or registration programs may address hazardous air pollutants
Requirements related to hydrogen sulfide gas	No specific requirements but CAA general duty clause requires prevention of accidental releases	6 of 6
Requirements related to flaring	Under new NSPS regulation, most hydraulically fractured gas wells must do green completions	6 of 6

Sources: GAO analysis of federal and state laws and regulations.

[a]Testing requirement applies only to certain wells—certain wells near proposed coalbed methane wells in Colorado and wells proposed for urbanized areas or in the vicinity of horizontal wells in Ohio. Pennsylvania does not require operators to identify or test nearby water wells, but state law incentivizes operators to do so by establishing a rebuttable presumption that operators are liable for changes in water quality of certain wells after drilling.

[b]Oil and gas well sites are only required to get permits for stormwater discharges if the facility has had a discharge of contaminated stormwater that includes a reportable quantity of a pollutant or contributes to the violation of a water quality standard, rather than prior to commencing construction or causing discharges.

[c]Generally, federal environmental laws do not have drilling, cementing, or casing requirements related to drilling production wells. However, according to EPA officials, if the well is to be hydraulically fractured with diesel fuel, it is subject to regulation as a Class II well under the SDWA UIC program and may be subject to cementing and casing, as well as plugging, requirements. In May 2012, EPA published draft guidance on how its UIC permit writers should address hydraulic fracturing with diesel in the context of the Class II UIC program. To date, however, EPA officials are unaware of any wells that were regulated in this way.

[d]Under TSCA, to the extent a hydraulic fracturing fluid is a chemical substance or mixture, manufacturers (including importers), processors, and distributors of such fluids generally would be subject to applicable reporting requirements. Generally, well site operators would not be subject to any such applicable TSCA reporting requirements.

[e]Disposal at a POTW is currently available in Pennsylvania and Ohio, as will be discussed later in this report. We are also aware that the city of Forth Worth, Texas had a pilot program within the last several years under which it accepted flowback for disposal through its POTW, but current information suggests that the city is no longer accepting flowback water.

[f]The existing federal regulations under RCRA solid waste provisions apply to nonhazardous waste disposal facilities and practices, including those involving oil and gas wastes, and prohibit open dumping of solid waste. However, EPA has a limited role in the enforcement of RCRA solid waste provisions.

Siting and Site Preparation

All six states we reviewed have state requirements regarding site selection and preparation, though the specifics of their requirements vary. Specifically, states have requirements for baseline testing of water wells, required setbacks from water sources, and stormwater management, among others. For example, three of the six states—Colorado, Ohio, and Pennsylvania—have requirements that encourage or require operators to conduct baseline water testing in certain cases. Colorado requires testing of certain nearby wells when a proposed coalbed methane well is located within a quarter-mile of a conventional gas well or a plugged and abandoned well.[72] In Ohio, baseline water well sampling is required within 1,500 feet of any proposed horizontal well or within 300 feet of any kind of well proposed in an urban area.[73] Pennsylvania does not require baseline testing, but state law presumes operators to be liable for any pollution of water wells within 2,500 feet of an unconventional well that occurs within 12 months of drilling activities, including hydraulic fracturing.[74] Operators in Pennsylvania can defend against this presumption if they have predrilling tests conducted by an independent certified laboratory showing that the pollution predated drilling. State regulators in Pennsylvania said that nearly all companies in Pennsylvania conduct baseline testing of nearby water wells, in many cases up to 4,000 feet from the drilling site.

Five of the six states—Colorado, North Dakota, Ohio, Pennsylvania, and Wyoming—we reviewed have requirements related to setbacks for well sites or equipment from certain water sources. For example, in Ohio, oil

[72]An abandoned well is a well that is no longer under control of an operator, whether or not it was properly plugged. In Colorado, wells must be tested for all major cations (positively-charged ions) and anions (negatively-charged ions), total dissolved solids, iron, manganese, selenium, nitrates and nitrites, dissolved methane, field pH, sodium adsorption ratio, presence of bacteria (iron related, sulfate reducing, slime, and coliform), specific conductance, and hydrogen sulfide.

[73]In Ohio, sampling must be conducted in accordance with state guidelines, which require testing for barium, calcium, iron, magnesium, potassium, sodium, chloride, conductivity, pH, sulfate, alkalinity, and total dissolved solids.

[74]Pennsylvania has a similar provision for conventional wells, which presumes operators to be liable for any pollution of water wells within 1,000 feet of a conventional well that occurs within 6 months of drilling activities.

and gas wells and associated storage tanks generally may not be within 50 feet of a stream, river, or other body of water. In Pennsylvania, unconventional wells may not be drilled within 500 feet of water wells without written owner consent unless the operator cannot otherwise access its mineral rights and demonstrates that additional protective measures will be utilized.[75] In Pennsylvania, there are also setbacks from public water supplies and certain other bodies of water such as springs and wetlands.

Oil and gas operations are generally not subject to certain stormwater permitting requirements under the Clean Water Act, but four of the six states we contacted—Colorado, North Dakota, Pennsylvania, and Wyoming—have their own stormwater permitting requirements. For example, the Wyoming Department of Environmental Quality requires permit coverage for stormwater discharges from all construction activities disturbing 1 or more acres. These permits require the operator to develop a stormwater management program, including best management practices, that can be reviewed by the Wyoming Department of Environmental Quality. In North Dakota, operators must obtain a permit for construction activities that disturb 5 or more acres, and state officials said that nearly all oil and gas drilling projects meet this threshold. This permit also requires the operator to develop a stormwater management program and implement best management practices for managing stormwater, such as using straw bales or dikes to manage water runoff. We did not identify any stormwater permitting requirements for Ohio and Texas, but their state regulations address stormwater in other ways. For example, operators in Ohio are required to comply with the state's best management practices during construction, such as design guidelines for constructing access roads. Texas regulations prohibit operators from causing or allowing pollution of surface water and encourage operators to implement best management practices to minimize discharges, including discharges of sediment during storm events.

States have additional requirements relating to erosion control, site preparation, and surface disturbance minimization. For more details about state siting and site preparation requirements, see appendix IX.

Drilling, Casing, and Cementing

All of the six states in our review have requirements related to how wells are to be drilled and casing should be installed and cemented in place,

---

[75]For conventional wells in Pennsylvania, the required setback is 200 feet.

though the specifics of their requirements vary. For example, states have different requirements regarding how deep operators must run surface casing to protect groundwater. In Pennsylvania, operators are required to run surface casing approximately 50 feet below the deepest fresh groundwater or at least 50 feet into consolidated rock, whichever is deeper. Generally, the surface casing may not be set more than 200 feet below the deepest fresh groundwater unless necessary to set the casing in consolidated rock.[76] Different casing and cementing requirements apply in Pennsylvania when drilling through coal formations, which state regulators said is common in the southwest part of the state. In Texas, operators are required to run surface casing to protect all usable quality water, as defined by the Texas Commission on Environmental Quality. The depth of the surface casing may be specified in a letter by the commission or in rules specific to a particular oil or gas field, which account for local considerations. In no case may surface casing be set deeper than 200 feet below the specified depth without prior approval from the Texas Railroad Commission, the oil and gas regulator in Texas. Operators in Wyoming are generally required to run surface casing to reach a depth below all known or reasonably estimated usable groundwater as defined in regulations and generally 100 to 120 feet below certain permitted water supply wells within a quarter-mile, but certain coalbed methane wells are exempt from these requirements. Until 2012, Ohio did not specify a depth to which surface casing was required to be set but according to state regulators, the depth of the casing used to protect groundwater was dictated through the permitting process, and regulators and operators were generally following the same casing and cementing requirements for unconventional wells as they would for Class II UIC wells. Ohio adopted new regulations effective August 2012 that generally require operators to run surface casing at least 50 feet below the base of the deepest underground source of drinking water or at least 50 feet into bedrock, whichever is deeper.

Among the six states we contacted, North Dakota and Ohio are the only states with specific casing and cementing provisions for horizontal wells. However, all six states have some requirements—whether through law, regulation, or the permitting process—that generally require operators to provide regulatory officials with information about the vertical and

---

[76]According to state regulators, this maximum surface casing depth requirement prevents fresh and brackish groundwater from commingling behind the same casing, among other things.

horizontal drilling paths. For example, an application for a permit to drill a horizontal well in Wyoming must include information about the vertical and horizontal paths of the well, and operators must provide notice to owners within a half-mile of any point on the entire length of the well. In addition, operators must (1) provide notification and obtain approval from the Wyoming Oil and Gas Conservation Commission before beginning horizontal drilling and (2) file a description of the exact path of the well, known as a directional survey, within 30 days of well completion. North Dakota requires a different permit to drill a horizontal well than it does for a vertical well, and the horizontal permit contains information about the horizontal path of the well.

For more details about state drilling, casing, and cementing requirements, see appendix IX.

Hydraulic Fracturing

All six states we reviewed have requirements for disclosing the chemicals used in hydraulic fracturing, but the specific requirements vary (see table 5). Four states—Colorado, North Dakota, Pennsylvania, and Texas—require disclosure through the website FracFocus, which is a joint project of the Ground Water Protection Council and the Interstate Oil and Gas Compact Commission.[77] For example, operators that perform hydraulic fracturing in Texas are required to upload certain information to the website FracFocus within 30 days after completion of the well or 90 days after the drilling operation is completed, whichever is earlier. Information required to be uploaded to FracFocus includes, among other things, the operator's name; the date of completion of hydraulic fracturing; the well location; the total volume of water used to conduct fracturing; and chemicals used, including their trade names, suppliers, intended use, and concentration. In Ohio, companies have options as to how to disclose information, including through FracFocus. Wyoming's chemical disclosure requirements were developed prior to the development of FracFocus, and the state does not require operators to disclose information through the website. Among the six states we contacted, Wyoming is the only state

---

[77]FracFocus is the national hydraulic fracturing chemical registry managed by the Ground Water Protection Council and Interstate Oil and Gas Compact Commission. The Ground Water Protection Council is a nonprofit organization whose members consist of state groundwater regulatory agencies, which come together to mutually work toward the protection of the nation's groundwater supplies. The Interstate Oil and Gas Compact Commission is a multistate government agency that works to ensure the nation's oil and natural gas resources are conserved and maximized while protecting health, safety, and the environment.

that requires operators to disclose certain chemical information prior to conducting hydraulic fracturing. Specifically, as part of their application for permit to drill, operators are required to submit information on the chemicals proposed to be used during hydraulic fracturing.

Table 5: Chemical Disclosure Requirements in Six Selected States

Requirements	Colorado	North Dakota	Ohio	Pennsylvania	Texas	Wyoming
Reporting mechanism	FracFocus website	FracFocus website	State website, FracFocus website, or other state approved method	FracFocus website	FracFocus website; state website if FracFocus is unavailable	State agency
Timing of disclosure requirement	Information must be disclosed within 60 days following completion of hydraulic fracturing and not later than 120 days after the commencement of hydraulic fracturing.	No timing requirement specified.	Information must be disclosed within 60 days after the completion of drilling or after a determination that a well is a dry or lost hole.	Information must be disclosed within 60 days following the conclusion of hydraulic fracturing.	Information must be disclosed within 30 days after completion of the well or within 90 days after drilling is completed, whichever is earlier.	Information must be disclosed before hydraulic fracturing and within 30 days of completion
Protections for confidential information or trade secrets	Yes	None identified[a]	Yes	Yes	Yes	Yes
Provisions to challenge trade secrets	None identified[a]	None identified[a]	Yes	None identified[a]	Yes	Yes
Provisions for disclosure to health professionals	Yes	None identified[a]	Yes	Yes	Yes	None identified[a]
Provisions for disclosure to emergency responders	Yes	None identified[a]	Yes	Yes	Yes	None identified[a]

Sources: GAO analysis of state requirements.

[a]We reviewed only those requirements specifically related to hydraulic fracturing disclosures. General state requirements related to the protection of confidential business information and protection of trade secrets may apply.

Five of the six states—Colorado, Ohio, Pennsylvania, Texas, and Wyoming—have specific provisions for protecting information on hydraulic fracturing fluids that is claimed as confidential business

GAO-12-874 Unconventional Oil and Gas Development

information or trade secrets. Four of the six states—Colorado, Ohio, Pennsylvania, and Texas—specifically require that the information must be provided to health professionals for diagnosis or treatment and to certain officials responding to a spill or a release. For example, in Texas, if an operator claims that a chemical is subject to trade secret protection, the chemical family or other similar description must generally be provided. Operators in Texas may not withhold information, including trade secrets, about chemicals used during hydraulic fracturing from health professionals or emergency responders who need the information for diagnostic, treatment, or other emergency response purposes, but health professionals and emergency responders must hold the information confidential except as required for sharing with other health professionals, emergency responders, or accredited laboratories for diagnostic or treatment purposes. Texas' regulations also allow for certain entities—including the owner of the land on which the well is located, an adjacent landowner, and relevant state agencies—to challenge a claim to trade secret protection.

Five of the six states—Colorado, North Dakota, Ohio, Pennsylvania, and Wyoming—have additional requirements specifically related to hydraulic fracturing. For example, Colorado, North Dakota, Ohio, and Wyoming require operators to continuously monitor certain pressure readings during hydraulic fracturing and to notify the state if pressure exceeds a certain threshold. Ohio also requires the suspension of operations when anticipated pressures are exceeded. North Dakota has mechanical integrity requirements specific to hydraulic fracturing, including requirements for specific types of casing, valves, and other equipment, which vary based on different fracturing scenarios. In addition, Colorado, Ohio, Pennsylvania, and Wyoming require operators to notify state regulators prior to conducting hydraulic fracturing, which provides state regulators the opportunity to conduct inspections during the hydraulic fracturing. Colorado requires notice 48 hours prior to conducting hydraulic fracturing, and Ohio and Pennsylvania require notice 24 hours prior. Wyoming does not require a specific period of notice. In Wyoming, benzene, toluene, ethylbenzene, and xylene (BTEX compounds) and petroleum distillates may only be used for hydraulic fracturing with prior authorization from state oil and gas regulators. Pennsylvania law requires blowout preventers to be used when drilling into an unconventional formation.[78]

For more details about state hydraulic fracturing requirements, see appendix IX.

Well Plugging

All six states in our review have requirements regarding well plugging, such as notifying the state prior to plugging or using specific materials or methods to do so. For example, operators in Colorado must obtain prior approval from state regulators for the plugging method and provide notice of the estimated time and date of plugging. Colorado regulations specify that the material used for plugging must be placed in the well in a manner that permanently prevents migration of oil, gas, water, or other substances out of the formation in which it originated. Cement plugs must be a minimum of 50 feet in length and must extend a minimum of 50 feet above each zone to be protected. After plugging the well, operators must submit reports of plugging and abandonment to the Colorado Oil and Gas Conservation Commission and include information specifying the fluid used to fill the wellbore, information about the cement used, date of work, and depth of plugs. In Pennsylvania, operators must follow (1) specific provisions for well plugging based on whether the well is located in a coal area or noncoal area or (2) an alternate approved method. Prior to plugging a well in an area underlain by a workable coal seam, the oil and gas operator must notify the state and the coal company to permit representatives to be present at the plugging.

In addition, all six states have programs to plug wells that were improperly plugged and have been abandoned, though their level of activity varies. For example, state regulators in Texas said that the primary objective of their program, which began in 1983, is to plug abandoned oil and gas wells that are causing pollution or threatening to cause pollution for which a responsible operator does not exist; the responsible operator failed to plug the well; or the responsible operator failed to otherwise bring the wells into compliance. As of 2009, Texas state regulators had plugged 30,000 wells, and approximately 8,000 potentially abandoned wells remained throughout the state. Officials stated, however, that many of

---

[78]Pennsylvania law defines an unconventional formation as a shale formation existing below a certain geologic interval where natural gas generally cannot be produced economically except by hydraulic fracturing or other specialized techniques. 58 Pa. Cons. Stat. § 2301 (2012).

these abandoned wells may be re-used for development of previously overlooked reservoirs. State regulators in North Dakota said that the number of abandoned wells in the state is very low compared with other states because the state was fairly late to oil and gas development—with major development starting in the 1950s—and that the state had a good tracking system in place during the early days of development. State regulators in North Dakota used funds from its well plugging program to plug two wells in the last year.

For more details about state well plugging requirements, see appendix IX.

Site Reclamation

All six states in our review have requirements for site reclamation, though the extent of the requirements varies. Five states—Colorado, Ohio, North Dakota, Pennsylvania, and Wyoming—have requirements both for backfilling soil and for revegetating areas. For example, in Colorado, final reclamation must generally be complete within 3 months of plugging a well on crop land and within 12 months on noncrop land. Reclamation in Colorado involves returning segregated soil horizons to their original relative positions;[79] returning crop land to its original contour; as near as practicable, returning noncrop land to its original contour to achieve erosion control and long-term stability; and adequately tilling to establish a proper seedbed. In Wyoming, operators must begin reclamation within 1 year of permanent abandonment of a well or last use of a pit and in accordance with the landowner's reasonable requests, or to resemble the original vegetation and contour of adjoining lands. In addition, where practical, topsoil must be stockpiled during construction for use in reclamation. Texas has requirements for contouring soil, but we did not identify requirements for revegetating the area.

For more details about state site reclamation requirements, see appendix IX.

Waste Management and Disposal

All six states in our review have some requirements regarding waste management and disposal, though specific requirements and practices vary across and within states. For example, regulators in Colorado said that the method of waste disposal varies based on the geological formation being exploited and the location of the production well. In some

---

[79]A soil horizon is a layer roughly parallel to the soil surface whose properties and characteristics differ from the layers above and beneath.

parts of the state, they said that the produced water generated is very salty and is therefore generally disposed of in a Class II UIC well. In contrast, in the Raton Basin—a coalbed methane formation near the border with New Mexico—the produced water is of sufficiently good quality that much of it is discharged to surface waters, according to state regulators.

All six states we reviewed have requirements regarding the use of pits for storage of produced water, drill cuttings, and other substances. For example, in North Dakota, a lined pit may be temporarily used to retain solids or fluids generated during activities including well completion, but the contents of the pits must be removed within 72 hours after operations have ceased and must be disposed of at an authorized facility. Pennsylvania requires that certain pits be lined and requires the liners to meet certain permeability, strength, thickness, and design standards; the pit itself must also be constructed so that it will not tear the liner and can bear the weight of the pit contents. In addition, Colorado and Wyoming require pitless drilling systems (tanks) to be used in certain circumstances. For example, Colorado requires pitless drilling systems for produced water from new oil and gas wells within a specified distance of certain drinking water supply areas, and Wyoming requires pitless drilling systems in areas where groundwater is less than 20 feet below the surface.

Underground injection of produced water in Class II UIC wells is a common method of disposal of produced water in five of the six states we reviewed.[80] For example, state regulators in Ohio said that there are 177 Class II UIC disposal wells currently in operation, and 98 percent of the fluid waste from oil and gas wells in Ohio is disposed of in these Class II UIC wells. As noted previously, five out of the six states we reviewed have primary responsibility for regulating injection wells, whereas EPA implements the program in Pennsylvania. The five states in our review that have been granted primacy for their Class II UIC programs obtained it under the alternative provisions in which they demonstrate to EPA that

---

[80]Pennsylvania has five currently operating Class II UIC disposal wells, and produced water generated in Pennsylvania is often recycled or shipped to other states such as Ohio for disposal. Until recently, EPA did not receive many applications for new Class II UIC wells in Pennsylvania. In the last 7 months, however, EPA officials said that they have received five permit applications for Class II UIC disposal wells and expect continued interest in the future.

Underground Injection and Earthquakes

From March 2011 to January 2012, 12 earthquakes ranging in magnitude from 2.1 to 4.0 occurred near Youngstown, Ohio. In March 2012, the Ohio Department of Natural Resources reported that "there is a compelling argument" that injection of produced water into nearby Class II UIC wells was the cause of the earthquakes. The Ohio Department of Natural Resources then placed a moratorium on injection into five Youngstown area UIC wells and began examining its Class II UIC well permitting process and developing a series of rule changes to help address seismic activity concerns. In July 2012, the Governor of Ohio signed an executive order determining that an emergency exists requiring the immediate adoption of the rule changes.

The National Academy of Sciences released a study in June 2012 that concluded underground injection does pose some risk for induced seismicity, but that very few events have been documented over the past several decades relative to the large number of disposal wells in operation. The study noted that the injected fluid volume, rate, pressure, and proximity to existing faults and fractures are factors that determine the probability to create a seismic event, but effective and economic tools are not currently available to accurately predict induced seismicity prior to injection. The study made research recommendations, proposed actions to address induced seismicity, and suggested that the agency that issues UIC well permits is the most appropriate agency to oversee decisions made with respect to induced seismic events.

their program is effective in preventing endangerment of underground sources of drinking water, in lieu of adopting all Class II UIC requirements in EPA regulations. All states have requirements for Class II UIC wells relating to casing and cementing, operating pressure, mechanical integrity testing, well plugging, and the monitoring and reporting of certain information, among other requirements. For example, North Dakota requires the operators of all new Class II UIC wells to demonstrate the mechanical integrity of the well and requires existing Class II UIC wells to demonstrate continued mechanical integrity at least once every 5 years. In North Dakota, mechanical integrity is demonstrated by showing that there is no significant leak in, for example, the casing; and there is no significant fluid movement into an underground source of drinking water through vertical channels adjacent to the injection well. Texas also requires operators to demonstrate the mechanical integrity of Class II UIC wells generally by conducting specified pressure tests before commencing injection, after conducting maintenance, and every 5 years. With regard to monitoring and reporting, Ohio requires operators to monitor injection pressures and volumes for each disposal well on a daily basis and to report annually on maximum and monthly average pressure and volumes.

Aside from underground injection, there are several other options for disposal of produced water, though the specifics vary across and within states. For example, regulatory agencies issue NPDES permits in Colorado, Texas, and Wyoming for direct discharges to surface waters in certain cases;[81] in doing so, the states must apply, where applicable, EPA's effluent limitations guidelines discussed above. According to state regulators in Wyoming, the state has about 1,000 currently active permits for discharges of produced water from coalbed methane formations and 500 permits for produced water from conventional formations. In contrast, state regulators in North Dakota said that there are no direct surface discharges of produced water in their state because the produced water is too salty.

Some states, such as Colorado and Pennsylvania, also have commercial facilities, which treat produced water before discharging it to surface waters. In addition, disposal to a POTW is an option in Ohio and

---

[81]Texas has not been authorized to issue NPDES permits for activities associated with the exploration, development, or production of oil or gas or geothermal resources. EPA is the NPDES permitting authority for those facilities in Texas.

Pennsylvania, but there have been some recent efforts to restrict such disposal.[82] One concern regarding disposal to POTWs is that these facilities may not have the technology necessary to remove key pollutants, including total dissolved solids, from the waste stream. In 2010, Ohio's Environmental Protection Agency (OEPA) approved a permit modification that allowed a POTW in Warren, Ohio, to accept 100,000 gallons per day of produced water with concentrations of less than 50,000 milligrams per liter of total dissolved solids, which was then diluted and discharged to surface waters.[83] However, the Director of OEPA subsequently issued a determination in 2011 that the permit had been unlawfully issued because Ohio law does not generally permit the disposal of produced water through a POTW.[84] In response, OEPA did not reauthorize the POTW to accept produced water when its NPDES permit came up for renewal in 2012. In July 2012, however, OEPA's decision was reversed by an administrative review commission, which held that the matter was outside of OEPA's jurisdiction. Instead, the power to prohibit disposal to a POTW lies with the Ohio Department of Natural Resources. Accordingly, the commission removed the NPDES permit's prohibition on accepting produced water.[85] Prior to 2011, POTWs in Pennsylvania also accepted produced water from oil and gas well sites. The Pennsylvania Department of Environmental Protection issued administrative orders to POTWs in Pennsylvania requiring, among other things, that the POTWs restrict the volume of oil and gas wastewater they were accepting, evaluate the impacts of oil and gas wastewaters on their treatment process, and submit certain samples of oil and gas wastewater

[82]As discussed earlier in this report, EPA sets pretreatment standards that apply when wastewater is sent to a facility—such as an industrial treatment facility or POTW—before being discharged to surface waters. To date, EPA has not set pretreatment standards specifically for produced water, though there is a general requirement that discharges to POTWs cannot cause the POTW to violate its own NPDES permit or interfere with treatment processes.

[83]The federal secondary standard for drinking water is 500 milligrams per liter.

[84]Ohio law provides that, generally, produced water must be disposed of only by underground injection, by surface application, in association with enhanced recovery of oil or gas resources from a well, or by other methods approved by the Chief of the Division of Oil and Gas Resources Management within the Ohio Department of Natural Resources for testing or implementing a new technology or method of disposal. Ohio Rev. Code Ann. § 1509.22(C)(1) (2012). According to OEPA officials, the permit did not involve an approved test or implementation of a new technology or method of disposal.

[85]Patriot Water Treatment, LLC v. Korleski, ERAC case Nos. 156477, 156588, 786501, and 786589 (2012).

accepted for treatment. In addition, the state of Pennsylvania requested that operators of Marcellus shale gas wells stop delivering produced water to POTWs and began revising the POTWs' NPDES permits. State officials later reported that POTWs in Pennsylvania were no longer accepting produced water from the Marcellus shale, and EPA Regional officials said that they believe that POTWs are accepting less produced water.

In addition to permanent disposal of produced water, all six states in our review allow for recycling or other reuses of produced water. For example, according to a 2011 report, over 50 percent of the produced water in Colorado is recycled.[86] In addition, state regulators in Pennsylvania said that the best option for dealing with produced water in the state is recycling, and the Department of Environmental Protection can track what percentage of recycled water was used in hydraulic fracturing based on information required on well completion reports. Approximately 90 percent of produced water in Pennsylvania is recycled, according to state regulators. The Texas Railroad Commission has approved several recycling projects in the Barnett Shale to reduce the amount of fresh water used in development activities there. Four of the six states—Colorado, North Dakota, Ohio, and Wyoming—also allow operators to reuse certain types of fluid waste for road applications. For example, in Ohio, produced water, excluding flowback from hydraulic fracturing, may be used for dust and ice suppression on roads with the approval of local governments; approximately 1 percent of produced water is used in this way. In Wyoming, road and land applications may be permitted as reuses of produced water. North Dakota allows road but not land application of produced water.

Regulatory agencies in all six states implement requirements for the disposal of waste such as drill cuttings. For example, in Colorado, drill cuttings may be buried in pits at the well site, an activity which is regulated by the Colorado Oil and Gas Conservation Commission. Drill cuttings taken off site for disposal at a commercial waste facility must comply with the regulations of the state's Department of Public Health and Environment that govern those facilities. Texas allows drill cuttings to be landfarmed on the well site where they were generated with the written permission of the

---

[86]State Review of Oil and Natural Gas Environmental Regulations, Inc., *Colorado Hydraulic Fracturing State Review* (Oklahoma City, OK: 2011).

surface owner of the site if they were obtained using drilling fluids with a chloride concentration of 3,000 milligrams per liter or less.[87] Texas allows on-site burial of drill cuttings that were obtained using drilling fluids with a chloride concentration in excess of 3,000 milligrams per liter. In North Dakota, operators frequently bury drill cuttings on-site where the North Dakota Industrial Commission's Oil and Gas Division has authority, but, in some cases, the drill cuttings may be disposed of at a landfill under the jurisdiction of the Department of Health due to shallow groundwater or permeable subsoil.

As discussed earlier in this report, officials in the six states we reviewed were not aware of any oil or gas well sites that would be regulated as small-quantity generators of hazardous waste under RCRA. Pursuant to RCRA, regulation of waste that is not considered hazardous is largely a state responsibility. Some states have special categories of waste and associated additional requirements that apply to industrial wastes generally, or oil and gas wastes specifically. For example, waste from crude oil and natural gas exploration and production in North Dakota is called special waste. Special waste landfills must be permitted and comply with specific design standards. Currently, there are four special waste landfills in North Dakota with another five proposed special waste landfills at the beginning stages of the permitting process. State regulators said that special waste consists mostly of drill cuttings but can also include other things such as contaminated soil. In Pennsylvania, oil and gas waste falls into a category of waste called residual waste that applies to, among other things, certain wastes from industrial, mining, or agricultural operations. Residual waste disposal must be permitted and is subject to processing and storage rules.

All six states in our review have requirements for managing and disposing of wastes, such as oilfield equipment, drilling solids, and produced water that have been exposed to or contaminated with naturally-occurring

---

[87]Landfarming is a method of treatment and disposal of low toxicity wastewater that involves spreading and mixing the wastewater into the soils to promote reduction of organic constituents and dilution and attenuation of metals. According to the Texas Railroad Commission, landfarming uses the physical, chemical, and biological capabilities of the soil-plant system to control waste migration and to provide a safe means of disposal without impairing the potential of the land for future use.

radioactive material (NORM) or technologically-enhanced NORM.[88] NORM occurs naturally in some geologic formations that also contain oil or gas and when NORM is brought to the surface during drilling and production, it remains in drill cuttings and produced water and, under certain conditions, creates scales or deposits on pipes or other oilfield equipment. Officials at the Colorado Department of Public Health and Environment said that they set tiers for how to manage materials that contain NORM based on their level of radioactivity. In addition, they said that the department is working with the Colorado Oil and Gas Conservation Commission to require operators to perform certain tests on produced water before allowing produced water to be used for road application. Texas officials said that the state requires operators to identify NORM-contaminated equipment with the letters "NORM" by securely attaching a clearly visible waterproof tag or marking with a legible waterproof paint or ink. In addition, Texas requires operators to dispose of oil and gas NORM waste by methods that are specifically authorized by rule or specifically permitted. State regulators in Wyoming said that a lot of NPDES permits for direct discharges to surface waters have limits on radioactivity that would probably lead the operator to dispose of produced water contaminated with NORM in a Class II UIC well.

For more details about states waste management and disposal requirements, see appendix IX.

Managing Air Emissions

Five of the six states we reviewed have permitting or registration requirements for managing air emissions from oil and gas production sites. In addition, all six states have requirements related to venting and flaring of gas and limiting or managing emissions of hydrogen sulfide—a hazardous and deadly gas—at drilling sites.

Five of the six states we reviewed —Colorado, North Dakota, Ohio, Texas, and Wyoming—have developed permitting or registration requirements that apply to oil and gas development. For example, according to state regulators, the vast majority of production wells in Colorado require air permits. Operators with certain condensate tanks and tank batteries are required to obtain a permit if the tanks have uncontrolled actual emissions of volatile organic compounds greater than or equal to 2 tons per year in areas which are not attaining certain air

---

[88]Technologically-enhanced NORM is produced when activities associated with oil and gas development concentrate or expose radioactive materials that occur naturally in soils, water, or other natural materials.

quality standards (nonattainment areas) or greater than or equal to 5 tons per year in an attainment area. As part of the permit requirements, operators in nonattainment areas must reduce emissions of volatile organic compounds by 90 percent from uncontrolled actual emissions during certain times of the year, and by 70 percent during other times, and reduce emissions by 90 percent for dehydration systems. In Ohio, an operator meeting certain requirements must obtain an air permit that lists each source of emissions; all applicable rules that apply to the sources, including federal and state requirements; operational restrictions; monitoring; recordkeeping; reporting; and testing requirements. Wyoming officials noted that oil and gas facilities are subject to general state permitting requirements but did not identify any permitting requirements specific to air emissions from oil and gas development. In Wyoming, state regulators have worked with industry to achieve voluntary reductions from mobile sources in certain parts of the state that may soon not meet air quality standards for ozone. Specifically, officials at the Wyoming Department of Environmental Quality said that they have asked operators in certain areas to agree to implement voluntary reductions in volatile organic compounds and nitrogen oxides and to install controls on diesel engines on mobile drilling rigs; regulators then include these requirements in the air permit issued to the operator. North Dakota and Texas also have permitting or registration requirements, and Pennsylvania is in the process of developing an inventory for oil and gas emissions information.

All six states have some requirements for flaring excess gas encountered during drilling and production, which may otherwise pose safety hazards and contribute to emissions. For example, operators in Pennsylvania who encounter excess gas during drilling or hydraulic fracturing must capture the excess gas, flare it, or divert it away from the drilling rig in a manner that does not create a hazard to public health and safety. According to state regulators in Wyoming, the Oil and Gas Conservation Commission has jurisdiction for flaring prior to production when the primary concern with flaring is safety. For flaring that occurs after production has begun, the Department of Environmental Quality requires 98 percent combustion efficiency.

All six states have safety requirements to limit and manage emissions of hydrogen sulfide—a hazardous and deadly gas—at drilling sites. For example, in Texas, operators are subject to detailed requirements in areas where exposure to hydrogen sulfide could exceed a certain threshold if a release occurred, taking into consideration whether the area of potential exposure includes any public areas such as roads. Requirements relate to posting warning signs, using fencing, maintaining protective breathing equipment at the well site, installing a flare line and a suitable method for lighting the flare, and conducting training. In some cases, hydrogen sulfide requirements overlap with flaring requirements. For example, flares used for treating gas containing hydrogen sulfide in North Dakota must be equipped and operated with an automatic ignitor or a continuous burning pilot, which must be maintained in good working order, including flares that are used for emergency purposes only.

For more details about state requirements for managing air emissions, see appendix IX.

Regional Commission Implements Additional Requirements

One of the states in our review—Pennsylvania—is also part of a regional commission that implements additional requirements governing several aspects of natural gas development. Specifically, the Delaware River Basin Commission is a regional body whose members include the governors of Delaware, New Jersey, New York, Pennsylvania, as well as the U.S. Army Corps of Engineers' Division Engineer for the North Atlantic Division. The commission regulates water quantity and quality within the basin, which spans approximately 13,500 square miles.[89]

In December 2010, the Delaware River Basin Commission published draft Natural Gas Development Regulations, which are currently under consideration for adoption, and the commission will not issue any permits for shale gas wells within the basin until the final regulations have been adopted. The draft regulations propose a number of requirements related to the protection of certain landscapes and waters and how to handle wastewater generated by natural gas development. For example, the proposed regulations require that produced water stored on the well pad be kept in enclosed tanks. In addition, operators of treatment and/or discharge facilities proposing to accept natural gas wastewater would be required to provide the commission with information on the contents of the proposed discharge and submit a study showing that the proposed discharge could be adequately treated. Natural gas well operators would also be required to have natural gas development plans for projects that exceed certain thresholds for acreage or number of wells. According to commission officials, the natural gas development plans would allow the

---

[89]The Susquehanna River Basin Commission—located partially in Pennsylvania—also regulates water withdrawals but not water quality in the context of oil or gas development and therefore was not included in our review.

commission to consider the cumulative impacts of development from numerous well pads, associated roads, and pipeline infrastructure, and to minimize and mitigate disturbance on lands most critical to water resources, such as core forests and steep slopes. The plans will also help protect water resources for approximately 15 million people, including residents of New York City and Philadelphia.

States Have Recently Updated Some Requirements

All six states in our review have updated some aspects of their requirements in recent years. Key examples include the following:

- Colorado made extensive amendments to its oil and gas regulations in 2008, which included, among other things, restrictions on locating wells near drinking water sources, measures to manage stormwater, and requirements to consult with the Colorado Division of Wildlife in certain cases to minimize adverse impacts on wildlife. According to state officials, these regulatory updates served three primary purposes: (1) address the growing impacts of increased oil and gas development; (2) implement state legislation passed in 2007 directing the Colorado Oil and Gas Conservation Commission to work with the Colorado Department of Public Health and Environment and the Colorado Division of Wildlife to update its regulations; and (3) update existing rules to enhance clarity, respond to new information, and reflect current practices and procedures.

- In 2012, North Dakota implemented 26 rule changes, including the requirement for operators to drain pits and properly dispose of their contents within 72 hours after well completion, servicing, or plugging operations have ceased. According to state officials, this change was implemented in response to a number of pit overflows that occurred during the spring melt in 2010 and 2011.

- In 2012, Ohio adopted new oil and gas well construction regulations to implement state legislation passed in 2010. The new regulations include casing and cementing requirements and requirements to disclose the chemicals used in hydraulic fracturing.

- Pennsylvania passed legislation in 2012 which, among other things, requires unconventional wells to be sited at greater setback distances from existing buildings and water wells than was previously required for all wells and requires chemical disclosure through FracFocus. In addition, the new legislation increases the distance from which an operator of an unconventional well may be presumed liable in the event of pollution of nearby water wells from 1,000 feet to 2,500 feet.

- The Texas Commission on Environmental Quality updated its air emissions regulations for oil and gas facilities in 2011, including emissions limitations for nitrogen oxide and volatile organic compounds. Texas officials told us that changes included requirements for operators to install controls on stationary compressor engines and storage tanks. In addition, operators in the Dallas-Fort Worth area have agreed to voluntarily reduce emissions of volatile organic compounds by replacing pneumatic valves with no-bleed or low-bleed valves which helps to address nonattainment issues in the area while also reducing emissions of hazardous air pollutants. Texas also adopted a regulation in December 2011 regarding chemical disclosure requirements in order to implement state legislation passed several months earlier.

- In 2010, Wyoming updated its chemical disclosure requirements. According to state regulators, operators were always required to provide notification to the Wyoming Oil and Gas Conservation Commission before conducting hydraulic fracturing, but recent regulatory changes clarified these requirements and also added detailed requirements on what information was required to be disclosed.

In the last 3 years, Colorado, Ohio, and Pennsylvania volunteered to have parts of their regulations reviewed by the State Review of Oil and Natural Gas Environmental Regulations (STRONGER) program, which is administered by the Ground Water Protection Council and brings together state, industry, and environmental stakeholders to review state oil and gas environmental regulations and make recommendations for improvement. Ohio and Pennsylvania have made regulatory changes that reflect STRONGER's recommendations. For example, STRONGER completed a review of Pennsylvania's regulations in September 2010. The review team commended the state for encouraging baseline groundwater testing in the vicinity of wells but also recommended that the state consider whether the testing radius should be expanded to take into account the horizontal portions of fractured wells. As discussed above, in 2012, Pennsylvania passed legislation that increases the distance from which an operator of an unconventional well may be presumed liable in the event of pollution of nearby water wells from 1,000 feet to 2,500 feet.

State regulators said that the addition was in response to the state's September 2010 STRONGER review and the Governor's Marcellus Shale Advisory Commission.[90] State regulators are also considering additional regulatory changes in response to the remaining recommendations of the Governor's Marcellus Shale Advisory Board.

Additional Requirements Apply on Federal Lands

Federal land management agencies, including the Bureau of Land Management (BLM), Forest Service, National Park Service, and Fish and Wildlife Service (FWS) manage federal lands for a variety of purposes. Specifically, both the Forest Service and BLM manage their lands for multiple uses, including oil and gas development; recreation; and provision of a sustained yield of renewable resources, such as timber, fish and wildlife, and forage for livestock. By contrast, the Park Service manages its lands to conserve the scenery, natural and historical objects, and wildlife so they remain unimpaired for the enjoyment of present and future generations. Similarly, FWS manages national wildlife refuges for the benefit of current and future generations, seeking to conserve and, where appropriate, restore fish, wildlife, plant resources, and their habitats.

Each of these agencies imposes additional requirements for oil and gas development on its lands to meet its obligations with respect to its mission. These additional federal requirements are the same for conventional and unconventional oil and gas development. In some cases, the surface rights to a piece of land and the right to extract oil and gas—called mineral rights—are owned by different parties. For example, private mineral rights might underlie lands where the surface is managed by a federal agency. Requirements for developing mineral rights vary based on whether the mineral rights are owned by the federal government or by a private entity.

---

[90]The purpose of the Pennsylvania Governor's Marcellus Shale Advisory Commission was to develop a comprehensive, strategic proposal for the responsible and environmentally sound development of the Marcellus Shale. Its membership consisted of the Lieutenant Governor, who served as the chair, and appointees chosen by the Governor and representing, among other things, the interests of environmental, conservation, industry, local and state government groups.

Requirements for Federally Owned Mineral Rights

Requirements for operators developing federally owned mineral rights are imposed by federal agencies during planning and leasing processes carried out by federal agencies. Operators must also meet specific requirements during several of the activities involved in oil and gas development.

Planning and Leasing Processes

BLM has primary authority for issuing leases and permits for federal oil and gas resources even in cases when surface lands are managed by other federal agencies or owned by private landowners. The majority of federal oil and gas leases underlie lands managed by BLM or the Forest Service, but there are some federal oil and gas resources available for leasing under lands managed by other federal agencies or private landowners.[91] Altogether, BLM oversees oil and gas development on approximately 700 million subsurface acres.

A first step in developing federal oil and gas resources is a planning phase, involving BLM and (for lands managed by the Forest Service) the Forest Service, to identify areas for potential leasing. Under the National Environmental Policy Act (NEPA), federal agencies are required to prepare a detailed statement on the environmental impacts of any "major federal action," if it would significantly affect the environment.[92] Regulations implementing NEPA generally require an agency to prepare

---

[91]BLM has also issued leases that underlie lands managed by the Park Service and FWS. These lands are generally not available for oil and gas development except in special circumstances. For example, FWS lands may be leased if an oil and gas operation outside of FWS lands is draining federal minerals under the FWS land. For the Park Service, small portions of three units of the National Park system are also open to federal mineral leasing based on their enabling legislation: Glen Canyon National Recreation Area, Lake Mead National Recreation Area, and Whiskeytown National Recreation Area. Currently, there are no parcels under lease in these areas. In order for oil and gas development to occur in these areas, the National Park system regional director must consent to the lease and permit, and can do so only upon determination that the activity permitted will not have significant adverse effect upon the resources or administration of the unit. (43 C.F.R. § 3109.2(b)). Three other units have a total of 16 wells under leases predating these policies.

[92]Pub. L. No. 91-190 (1970), codified as amended at 42 U.S.C. §§ 4321-4347 (2012).

either an environmental assessment or environmental impact statement.[93] After the planning process, BLM takes the lead in preparing the NEPA analysis for leases when the surface lands are managed by BLM or owned by a private landowner (see table 6). For Forest Service lands, the Forest Service takes the lead in preparing the NEPA analysis and coordinates with BLM so that BLM's subsequent leasing decision can be supported by the same analysis. At both agencies the NEPA review focuses on how the sale of leases may affect the environment and public health and, according to BLM officials, often includes mitigation measures that ultimately become stipulations on leases and permits for that tract of federal land. After the environmental impact statement is completed, BLM sells the lease to an operator through an auction or by other means.

Table 6: Surface Agency Roles in Leasing and Permitting Federal Minerals

Surface land management agency	Availability for oil and gas development	Role in approving leases	Role in approving drilling permits
BLM	Generally available[a]	BLM is primary authority for approving leases	BLM is primary authority for approving drilling permits
Forest Service	Generally available[b]	Coordinates with BLM regarding surface issues and must authorize the lease	Forest Service must approve all surface disturbing activities before BLM approves the drilling permit.
Private Landowner	Generally available	No specific role; may participate in public comment process	Operator generally must reach agreement with surface landowner regarding surface disturbances.

Source: GAO.

[a]Some lands managed by BLM are unavailable for leasing because they have been withdrawn from leasing through congressional action or by agency regulation or, according to BLM officials, because they are being managed for other uses.

[93]Agencies may prepare an environmental assessment—a concise public document—that provides sufficient evidence and analysis for determining whether to prepare an environmental impact statement or a finding of no significant impact. An environmental impact statement is a more detailed statement than an environmental assessment, and NEPA implementing regulations specify requirements and procedures—such as providing the public with an opportunity to comment on the draft document. An environmental impact statement must, among other things, (1) describe the environment that will be affected, (2) identify alternatives to the proposed action and identify the agency's preferred alternative, (3) present the environmental impacts of the proposed action and alternatives, and (4) identify any adverse environmental impacts that cannot be avoided should the proposed action be implemented. The environmental impact statement or environmental assessment may also be used to document that the proposed leasing action is in compliance with other federal laws, including the National Historic Preservation Act (intended to preserve historic and archeological sites) and the Endangered Species Act (intended to protect threatened and endangered species).

ᵇSome lands managed by the Forest Service are unavailable for leasing because they have been withdrawn through congressional action or by agency regulation. For example, the Wyoming Range (1.2 million acres) in western Wyoming and the Valle Vidal (100,000 acres) in New Mexico are both unavailable for leasing. In addition, 58.5 million acres of Inventoried Roadless Areas are indirectly unavailable in that construction or reconstruction of roads is not allowed, essentially making these areas inaccessible, according to Forest Service officials.

After acquiring a lease for the development of federal oil and gas, an operator is required to submit an application for permit to drill (APD) for individual wells to BLM. According to BLM officials, the APD is a comprehensive plan for drilling and related activities, which is approved by BLM. Prior to permit issuance for the proposed drilling activity, BLM is required to document that needed reviews under NEPA have been conducted. According to officials, at this step BLM conducts site-specific NEPA analysis, often drawing on the previous NEPA analysis conducted prior to the lease sale, but supplemented with more specifics about the proposed well site and related facilities, such as access roads or pipelines. The environmental review may also identify mitigation measures that could be used to reduce the environmental effects of drilling. The APD includes two key components: (1) the drilling plan, which describes the plan for drilling, casing, and cementing the well; and (2) the surface use plan of operations, which describes surface disturbances, such as road construction to the well pad and installation of any needed pipelines or other infrastructure. BLM is responsible for reviewing and approving the APD as a whole but gets input from the surface land management agency regarding the surface use plan of operations. For example, the Forest Service is responsible for review and approval of the surface use plan of operations component of the APD. After reviewing the operator's APD, BLM approves the APD, often by attaching conditions of approval and requiring the operator to take mitigation measures as described in the environmental review or recommended by the surface land management agency. Once the APD is approved, and any state or local approvals are obtained, the operator can begin work.

BLM has overall responsibility for ensuring compliance with approved APDs but coordinates with other surface land management agencies as appropriate. According to BLM officials, BLM is responsible for inspections and enforcement related to drilling operations, including running tests on casing and cementing. In addition, BLM officials said that they coordinate with surface land management agencies regarding surface conditions. Forest Service officials said that the Forest Service is responsible for conducting inspections relative to surface uses authorized by the surface use plan of operations. These officials said that if Forest

Service personnel note possible noncompliance related to drilling or production operations, they notify and coordinate with BLM. Similarly, officials said that, if BLM conducts an inspection and notices potential violations of the surface use plan of operations, they contact the Forest Service.

Requirements Related to Oil and Gas Development Activities

Operators of wells accessing federal oil and gas also face requirements related to activities involved in oil and gas development. Specifically, these requirements are related to siting and site preparation; drilling, casing, and cementing; well plugging; site reclamation; waste management and disposal; and managing air emissions. Requirements are as follows:

- *Siting and site preparation.* BLM requires an operator to identify all known oil and gas wells within a 1-mile radius of the proposed location. BLM does not require baseline testing of groundwater near the proposed well site. BLM generally prohibits an operator from conducting operations in areas highly susceptible to erosion, such as floodplains or wetlands, and recommends that operators avoid steep slopes and consider temporarily suspending operations when weather-related conditions, such as freezing or thawing ground, would cause excessive impacts.

- *Drilling, casing, and cementing.* As discussed above, operators must submit detailed drilling plans as part of their APD. The drilling plan must be sufficiently detailed for BLM to appraise the technical adequacy of the proposed project and must include, among other things: (1) geologic information about the formations that the operator expects to encounter while drilling; (2) whether these formations contain oil, gas, or useable water and, if so, how the operator plans to protect such resources; (3) a proposed casing plan, including details about the size of the casing and the depths at which each layer of casing will be set; (4) the estimated amount and type of cement to be used in the well; and (5) a description of any horizontal drilling that is planned.

- *Well plugging.* Operators are required to provide notice to and get approval from BLM prior to plugging a well and to comply with specific technical standards in plugging the well.

- *Site reclamation.* Operators describe their plans for reclamation in the surface use plan of operations submitted as part of the APD. BLM requires operators to return the disturbed land to productive use. All well pads, pits, and roads must be reclaimed and revegetated. Interim and final reclamation generally must be completed within 6 months of the well entering production and being plugged, respectively.

- *Waste management and disposal.* In the surface use plan of operations, operators must describe the methods and locations proposed for safe disposal of wastes, such as drill cuttings, salts, or chemicals that result from drilling the proposed well. The description must also include plans for the final disposition of drilling fluids and any produced water recovered from the well.

- *Managing air emissions.* For operations in formations that could contain hydrogen sulfide, BLM requires a hydrogen sulfide operations drilling plan, which describes safety systems that will be used, such as detection and monitoring equipment, flares, and protective equipment for essential personnel.[94]

In some cases, BLM and states may regulate similar activities; in such cases, operators must comply with the more stringent regulation. For example, North Dakota state requirements allow the use of pits only for short-term storage of produced water. BLM generally allows the use of pits for longer-term storage of produced water, but operators cannot do so on federal lands in North Dakota due to state requirements. See appendix X for a comparison of federal environmental requirements, state requirements, and additional requirements that apply on federal lands.

BLM recently proposed new requirements for oil and gas development on federal lands. Specifically, in May 2012, BLM proposed regulations that update and add to its current requirements related to hydraulic fracturing. As proposed, these regulations would require operators of wells under federal leases to (1) publicly disclose the chemicals they use in hydraulic fracturing; (2) take certain steps to ensure the integrity of the well, including complying with certain cementing standards and confirming through mechanical integrity testing that wells to be hydraulically fractured meet appropriate construction standards; and (3) develop plans for managing produced water from hydraulic fracturing and store flowback water from hydraulic fracturing in a lined pit or a tank. According to BLM officials, BLM's proposed rule is intended to improve stewardship and operational efficiency by establishing a uniform set of standards for

---

[94]BLM and the Forest Service also include air quality impacts in the NEPA analysis conducted for leasing and permitting actions.

hydraulic fracturing on public lands. According to BLM officials, a final rule is expected in the fall of 2012.

Requirements for Privately Owned Mineral Rights under Federal Surface Lands

Subject to some restriction, owners of mineral rights that underlie federal lands have the legal authority to explore for oil and gas and, if such resources are found, to develop them.[95] Federal land management agencies' authorities to control the surface impacts of drilling for privately owned minerals underlying federal lands vary based on a variety of factors, including which federal agency is responsible for managing the surface lands.[96]

According to BLM officials, private mineral owners seeking to develop oil and gas would need to obtain a right-of-way grant from BLM for any surface disturbance, including the well pad, but otherwise BLM has limited authority over the private owners' use and occupancy of the BLM-managed surface lands. Officials said that BLM would have the same rights as a private surface owner under state law to hold a mineral rights owner to "reasonable surface use." BLM officials explained that BLM would perform a NEPA analysis prior to issuing the right-of-way grant. According to officials, the agency applies its general regulations for granting rights of way, but BLM did not have specific guidance regarding oversight of private mineral operations on BLM lands.

According to Forest Service officials, Forest Service authority related to the development of privately owned minerals is limited because private mineral owners have the legal right to develop such resources. The

---

[95]In implementing requirements on the development of private mineral rights, agencies must consider the potential applicability of the Fifth Amendment to the U.S. Constitution. The Fifth Amendment prohibits the federal government from taking private property for public use without justly compensating the private property owner. Government regulation may place restrictions on the use of property to the extent that it deprives the owner of its use or economic value. In such cases of "regulatory taking," the owner may be entitled to just compensation under the Fifth Amendment. Thus, if agency requirements "regulated" the mineral rights to the point that they were deemed to be taken, the agency would have to compensate the owner. See, e.g., *Foster v. United States*, 607 F.2d 943 (Ct. Cl. 1979) (government's refusal to allow permit holders of mineral interest on government land any right of access for the purpose of extracting minerals was a compensable taking).

[96]In addition, an agency's authority may vary depending on whether the private rights were severed from the surface land before the land was conveyed to the United States, or were retained by the owner who conveyed the land to the United States; whether the lands in question are public domain or acquired; and on other factors.

GAO-12-874 Unconventional Oil and Gas Development

Forest Service manages a large number of wells accessing privately owned minerals. Specifically, Forest Service officials said that, of the 19,000 operating oil and gas wells on Forest Service lands, about three-fourths are producing privately owned minerals.[97] Forest Service officials explained that the Forest Service evaluates the effects of the development and, through negotiations with the operator, tries to reach agreement on certain mitigation measures. Officials explained that these mitigation measures are generally not as stringent or specific as mitigation measures used on federal leases. In addition, Forest Service officials explained that enforcement options are limited for environmental damage from development of privately owned minerals. Generally, the Forest Service can work with state oil and gas agencies to have them enforce any relevant state requirements regarding surface impacts, or the Forest Service can seek an injunction from the court to stop damaging actions and then pursue possible damages or restitution via the court. According to Forest Service officials, development of privately owned minerals has been a particular challenge in the Alleghany National Forest in Pennsylvania where privately owned minerals underlie more than 90 percent of the forest. Forest Service officials stated that there are approximately 1,000 new wells drilled in this forest each year, most of which are shallow conventional oil development. Officials said that the pace of this development has made it difficult for the Forest Service to manage other forest uses, such as recreation and timber extraction.

Regarding lands managed by FWS, we reported in August 2003 that oversight and management of oil and gas activities varies widely among wildlife refuges.[98] We noted that some refuges issue permits that establish operating conditions for oil and gas activities, which give these refuges greater control over oil and gas activities and protect refuge resources; other refuges exercise little control or enforcement over oil and gas activities. According to FWS officials, this situation persists today,

---

[97]According to Forest Service officials, private mineral ownership is more common in the eastern United States because most eastern forest lands were acquired through the 1911 Weeks Act, which had different stipulations for mineral rights of lands conveyed to the Forest Service. See also *Minard Run v. U.S. Forest Service*, 2009 WL 4937785 at *3 (W.D.Pa.) (discussing the Weeks Act); and GAO, *Private Mineral Rights Complicate the Management of Eastern Wilderness Areas*, GAO/RCED-84-101 (Washington, D.C.: July 26, 1984).

[98]See GAO, *National Wildlife Refuges: Opportunities to Improve the Management and Oversight of Oil and Gas Activities on Federal Lands*, GAO-03-517 (Washington, D.C.: Aug. 28, 2003).

GAO-12-874 Unconventional Oil and Gas Development

partly because FWS does not currently have regulations that directly address oil and gas development. FWS officials said that the agency is developing a proposed rule that will set requirements for operators developing privately owned minerals. Officials expect an Advance Notice of Proposed Rulemaking to be issued in calendar year 2012. FWS officials said that, despite having minimal requirements for operators drilling for privately owned minerals, they can use other federal authorities and work with federal and state agencies to minimize or remediate injury to FWS lands.[99] For example, FWS worked with EPA to respond to a spill of produced water into a stream on a National Wildlife Refuge in Louisiana in 2005, in violation of CWA. EPA, the Coast Guard, and the Department of Justice worked together on the case, and the operator ultimately paid $425,000 to FWS for the two affected wildlife refuges. According to agency officials, however, without specific regulations, FWS faces challenges conducting daily management and oversight of oil and gas activities on FWS lands.

The Park Service's 9B regulations govern potential impacts to all park system resources and values resulting from exercise of private oil and gas rights within Park Service administered lands. These regulations require an operator to submit a proposed plan of operations to the Park Service, which outlines the activities that are proposed for Park Service lands, including drilling, production, transportation, and reclamation. The regulations also outline certain requirements for operators, including that operations be located at least 500 feet from surface waters, that fences be used to protect people and wildlife, and that during reclamation the operator reestablish native vegetation. The Park Service analyzes the operator's proposed plan of operations to ensure that the proposed plan complies with the 9B regulations. Also, in determining whether it can approve an operation, the Park Service undertakes an environmental analysis under NEPA. Once the Park Service approves the proposed plan of operations, the operator can begin drilling. The Park Service continues to have access to the site for monitoring and enforcement purposes. In November 2009, the Park Service issued an Advance Notice of Proposed Rulemaking to update its 9B regulations; a proposed rule is expected in September 2013, according to agency officials.

---

[99]Officials said that these other federal authorities could include, for example, CWA, Endangered Species Act, or Migratory Bird Conservation Act.

Federal and State Agencies Reported Several Challenges Regulating Unconventional Oil and Gas Development

Federal and state agencies reported facing several challenges in regulating oil and gas development from unconventional reservoirs. Specifically, EPA officials reported that their ability to conduct inspection and enforcement activities and limited legal authorities are challenges. In addition, BLM and state officials reported that hiring and retaining staff and educating the public are challenges.

Conducting Inspection and Enforcement Activities

Officials at EPA reported that conducting inspection and enforcement activities for oil and gas development from unconventional reservoirs is challenging due to limited information, as well as the dispersed nature of the industry and the rapid pace of development. More specifically, according to EPA headquarters officials, enforcement efforts can be hindered by a lack of information in a number of areas. For example, in cases of alleged groundwater contamination, EPA would need to link changes in groundwater quality to oil and gas activities before taking enforcement actions. However, EPA officials said that often no baseline data exist on the quality of the groundwater prior to oil and gas development.[100] These officials also said that linking groundwater contamination to a specific activity may be difficult even in cases where baseline data are available because of the variability and complexity of geological formations.

In addition, EPA officials said that they do not always have information on the types of activities taking place or equipment being used at oil and gas well sites, making it difficult to know where to conduct inspections related to SDWA, CWA, and CAA. For example, regarding SDWA, EPA headquarters officials said that, though EPA's guidance document on this topic is not yet finalized, EPA requires operators conducting hydraulic fracturing operations with diesel fuel to apply for a Class II UIC permit.[101] However, it is difficult for EPA to assess operators' compliance because

---

[100]As discussed earlier in this report, three of the six states we reviewed—Colorado, Ohio, and Pennsylvania—have requirements that encourage or require operators to conduct baseline water testing in certain cases.

[101]As discussed earlier in this report, in 2005, the Energy Policy Act amended SDWA to specifically exempt hydraulic fracturing from the UIC program, unless diesel fuel is used in the hydraulic fracturing process.

GAO-12-874 Unconventional Oil and Gas Development

the agency does not know which operators are using diesel. Similarly, with respect to CWA, EPA officials said it is difficult to assess operators' compliance with the SPCC program, which establishes spill prevention and response planning requirements in accordance with CWA, because EPA does not know the universe of operators with tanks subject to the SPCC rule. In addition, related to CAA, EPA headquarters officials said that it would be difficult for EPA to find oil and gas wells that are subject to but noncompliant with NESHAPs because EPA does not have information on the universe of oil and gas well sites with the equipment that are significant to air emissions. Also, according to EPA Region 8 officials, these requirements are "self-implementing," and EPA would only receive notice from a facility that identifies itself as subject to the rules.

Several EPA officials also mentioned that the dispersed nature of the industry and the rapid pace of development make conducting inspections and enforcement activities difficult. For example, officials in EPA Region 5 said that it is a challenge to locate the large number of new well sites across Ohio and to get inspectors out to these sites because EPA generally does not receive information about new wells or their location.[102] EPA headquarters officials also mentioned that many oil and gas production sites are not continuously staffed, so EPA needs to contact operators and ensure that someone will be present before visiting a site to conduct an inspection. Officials in EPA Region 6 said that the dispersed nature of the industry, the high level of oil and gas development in the Region, and the cost of travel have made it difficult to conduct enforcement activities in their Region.

EPA officials in headquarters said that SDWA is a difficult statute to enforce because of the variation across states. Specifically, SDWA authorizes EPA to approve, for states that elect to assume this responsibility, individual states' programs as alternatives to the federal UIC Class II regulatory program. As a result, EPA's enforcement actions have to be specific to each state's program, which increases the complexity for EPA. In addition, SDWA requires that EPA approve each state's UIC program by regulation rather than through an administrative process, and many of the federal regulations for state UIC programs are out of date. EPA officials said that this has hindered enforcement efforts, and some cases have been abandoned because EPA can only enforce

---

[102]EPA Region 5 includes Indiana, Illinois, Michigan, Minnesota, Ohio, and Wisconsin.

those aspects of state UIC regulations that have been approved by federal regulation.

Limited Legal Authorities

EPA officials also reported that the scope of their legal authorities for regulating oil and gas development is a challenge. For example, EPA officials in headquarters and Regional offices told us that the exclusion of exploration and production waste from hazardous waste regulations under RCRA significantly limits EPA's role in regulating these wastes. For example, if a hazardous waste permit was required, then EPA would obtain information on the location of well sites, how much hazardous waste is generated at each site, and how the waste is disposed of; however, operators are not required to obtain hazardous waste permits for oil and gas exploration and production wastes, limiting EPA's role.[103] As discussed earlier in this report, EPA is currently considering a petition to revisit the 1988 determination not to regulate these wastes as hazardous, but according to officials, has no specific time frame for responding. In addition, as we described earlier in this report, officials in Region 8 noted that EPA cannot use either its CERCLA or CWA emergency response authority to respond to spills of oil if there is no threat to U.S. navigable waters or adjoining shorelines because those statutory authorities do not extend to such situations.[104]

Hiring and Retaining Staff

Officials at BLM, Forest Service, and state agencies reported challenges hiring and retaining staff. For example, BLM officials in North Dakota said recruiting is a challenge because the BLM pay scale is relatively low compared with the current cost of living near the oil fields in the Bakken formation. Similarly, BLM officials in North Dakota and headquarters both said that retaining employees is difficult because qualified staff are frequently offered more money for private sector positions within the oil and gas industry. BLM officials in Wyoming told us that their challenges related to hiring and retaining staff have made it difficult for the agency to keep up with the large number of permit requests and meet certain inspection requirements. We previously reported that BLM has

---

[103]EPA officials said that, at present, they could specifically request this type of information, but do not receive it automatically.

[104]EPA Region 8 includes Colorado, Montana, North Dakota, South Dakota, Utah, and Wyoming.

encountered persistent problems in hiring, training, and retaining sufficient staff to meet its oversight and management responsibilities for oil and gas operations on federal lands. For example, in March 2010, we reported that BLM experienced high turnover rates in key oil and gas inspection and engineering positions responsible for production verification activities.[105] We made a number of recommendations to address this and other issues—and the agency agreed—but we reported in 2011 that the human capital issues we identified with BLM's management of onshore oil and gas continue.[106]

State oil and gas regulators in two of the six states we reviewed—North Dakota and Texas—also reported challenges with employees leaving their agencies for higher paying jobs in the private sector. Officials from the North Dakota Industrial Commission—which regulates oil and gas development—said they have partially mitigated this challenge by removing state geologists and engineers from the traditional state pay scale and offering signing and retention bonuses. In addition, state environmental regulators in three of the six states—North Dakota, Pennsylvania, and Wyoming—also mentioned challenges related to hiring or retaining staff. For example, air regulators in the Wyoming Department of Environmental Quality said that retaining qualified staff is challenging, as staff leave for higher-paying private sector positions. These officials said that 6 of their 22 air permit-writing positions are vacant as of June 2012. State regulators in Colorado and Ohio did not report facing this challenge.

In addition, FWS officials reported that they have inadequate staffing for oil and gas development issues and noted that additional regional and field positions could help FWS implement a more comprehensive oil and gas program.

Public Education

BLM and state officials reported that providing information and education to the public is a challenge. Specifically, BLM headquarters officials mentioned that hydraulic fracturing has attracted the interest of the public

---

[105]GAO, *Oil and Gas Management: Interior's Oil and Gas Production Verification Efforts Do Not Provide Reasonable Assurance of Accurate Measurement of Production Volumes*, GAO-10-313 (Washington, D.C.: Mar. 15, 2010).

[106]GAO, *Oil and Gas Leasing: Past Work Identifies Numerous Challenges with Interior's Oversight*, GAO-11-487T (Washington, D.C.: Mar. 17, 2011).

and that BLM has been fielding many information requests about its use in oil and gas development. In addition, officials in five of the six states—Colorado, Ohio, Pennsylvania, Texas, and Wyoming—reported challenges related to public education. For example, regulators in Ohio said that their agency has conducted more public outreach in the last year than in the past 20 years and, in response to this public interest in shale drilling and hydraulic fracturing, they will be adding more communications staff. Similarly, oil and gas development is moving into areas of Colorado that are not accustomed to this development, and state officials in both the Department of Public Health and Environment and the Oil and Gas Conservation Commission said that they have spent a lot of time providing the public with information on topics including hydraulic fracturing. State regulators in Wyoming said that educating the public has been a challenge since coalbed methane and tight sandstone development in Wyoming is very different than, for example, shale gas development in Pennsylvania, but the media do not always make this clear. State regulators in North Dakota did not report public education as a challenge.

Agency Comments and Our Evaluation

We provided a draft of this report to EPA and to the Departments of Agriculture and the Interior for review and comment. The Departments of Agriculture and Interior provided written comments on the draft, which are summarized below and appear in their entirety in appendixes XI and XII, respectively. In addition, both Departments and EPA provided technical comments, which we incorporated as appropriate.

In its written comments, the Department of Agriculture agreed with our findings and noted that the Forest Service also faces challenges hiring and retaining qualified staff. In response, we added this information to the report.

In its written comments, the Department of the Interior provided additional clarifying information on its efforts concerning BLM's proposed rule on hydraulic fracturing and steps BLM is taking to hire and retain skilled technical staff. In response, we included additional information in the report about BLM's proposed rule on hydraulic fracturing.

As agreed with your offices, unless you publicly announce the contents of this report earlier, we plan no further distribution until 30 days from the report date. At that time, we will send copies of this report to the appropriate congressional committees, the EPA Administrator, the Secretaries of Agriculture and the Interior, the Director of the Bureau of Land Management, and other interested parties. In addition, this report will be available at no charge on the GAO website at http://www.gao.gov.

If you or your staff members have any questions about this report, please contact me at (202) 512-3841 or trimbled@gao.gov. Contact points for our Offices of Congressional Relations and Public Affairs may be found on the last page of this report. GAO staff who made major contributions to this report are listed in appendix XIII.

David C. Trimble
Director
Natural Resources and Environment

List of Requesters

The Honorable Barbara Boxer
Chairman
Committee on Environment and Public Works
United States Senate

The Honorable Sheldon Whitehouse
Chairman
Subcommittee on Oversight
Committee on Environment and Public Works
United States Senate

The Honorable Benjamin L. Cardin
Chairman
Subcommittee on Water and Wildlife
Committee on Environment and Public Works
United States Senate

The Honorable Henry A. Waxman
Ranking Member
Committee on Energy and Commerce
House of Representatives

The Honorable Edward J. Markey
Ranking Member
Committee on Natural Resources
House of Representatives

The Honorable Diana DeGette
Ranking Member
Subcommittee on Oversight and Investigations
Committee on Energy and Commerce
House of Representatives

The Honorable Robert P. Casey, Jr.
United States Senate

Appendix I: Objectives, Scope, and Methodology

To identify federal and state environmental and public health requirements governing onshore oil and gas development from unconventional reservoirs, we analyzed federal and state laws, regulations, and guidance, as well as reports on federal and state requirements. We defined unconventional reservoirs as including shale gas deposits, shale oil, coalbed methane, and tight sandstone formations. We focused our analysis on requirements that apply to activities on the well pad and wastes or emissions generated at the well pad rather than on downstream infrastructure such as pipelines or refineries. In particular, we identified and reviewed eight key federal environmental and public health laws, specifically the Safe Drinking Water Act; Clean Water Act; Clean Air Act; Resource Conservation and Recovery Act; Comprehensive Environmental Response, Compensation, and Liability Act; Emergency Planning and Community Right-to-Know Act; Toxic Substances Control Act; and Federal Insecticide, Fungicide, and Rodenticide Act. We also reviewed corresponding regulations such as the Environmental Protection Agency's (EPA) *New Source Performance Standards and National Emission Standards for Hazardous Air Pollutants for the Oil and Gas Industry* and guidance such as EPA's *Guidance for Implementation of the General Duty Clause of the Clean Air Act.*

To identify state requirements, we identified and reviewed laws and regulations in a nonprobability sample of six selected states—Colorado, North Dakota, Ohio, Pennsylvania, Texas and Wyoming. We selected states with current unconventional oil or gas development and large reservoirs of unconventional oil or gas. In addition, we ensured that the selected states included a variety of types of unconventional reservoirs, differing historical experiences with the oil and gas industry, and that some of the selected states have significant oil and gas development on federal lands. Because we used a nonprobability sample, the information that we collected from those states cannot be generalized to all states but can provide illustrative examples.

To complement our analysis of federal and state laws and regulations, we interviewed officials in federal and state agencies to discuss how federal and state requirements apply to the oil and gas industry (see table 7). In particular, we interviewed officials in EPA headquarters and four Regional offices where officials are responsible for implementing and enforcing programs within the six states we selected, including Region 3 for Pennsylvania, Region 5 for Ohio, Region 6 for Texas, and Region 8 for Colorado, North Dakota, and Wyoming. We also interviewed state officials responsible for implementing and enforcing requirements governing the oil and gas industry and environmental or public health requirements in each of the six states we selected. For three of these states—Colorado, North Dakota, and Wyoming—we conducted these interviews in person. We also interviewed officials from the Delaware River Basin Commission—a regional body that manages and regulates certain water resources in four states, including Pennsylvania. We also contacted officials from environmental, public health, and industry organizations to gain their perspectives and to learn about ongoing litigation or petitions that may impact the regulatory framework. We selected environmental organizations that had made public statements about federal or state requirements for oil and gas development and public health organizations representing state and local health officials and communities. The selected organizations are a nonprobability sample, and their responses are not generalizable. In addition, we visited drilling, hydraulic fracturing, and production sites in Pennsylvania and North Dakota and met with company officials to gather information about these processes and how they are regulated at the federal and state levels. We selected these companies based on their operations in the six states we selected.

Table 7: Agencies and Organizations Contacted

Federal agencies

• EPA Office of Air and Radiation	• Bureau of Land Management (BLM) headquarters
• EPA Office of Chemical Safety and Pollution Prevention	• BLM Colorado State Office
• EPA Office of Enforcement and Compliance Assurance	• BLM Dickinson, North Dakota Field Office
• EPA Office of General Counsel	• BLM Wyoming State Office
• EPA Office of the Inspector General	• Fish and Wildlife Service
• EPA Office of Solid Waste and Emergency Response	• Forest Service
• EPA Office of Research and Development	• National Park Service
• EPA Office of Water	
• EPA Region 3 (includes Pennsylvania)	
• EPA Region 5 (includes Ohio)	
• EPA Region 6 (includes Texas)	
• EPA Region 8 (includes Colorado, North Dakota, and Wyoming)	

State and regional agencies

- Colorado Oil and Gas Conservation Commission
- Colorado Department of Public Health and Environment
- Air Pollution Control Division
- Hazardous Materials and Waste Management Division
- Water Quality Control Division
- Delaware River Basin Commission
- Ground Water Protection Council
- North Dakota Industrial Commission, Oil and Gas Division
- North Dakota Department of Health, Environmental Health Section
- Air Quality Division
- Waste Management Division
- Municipal Facilities Division
- Ohio Department of Natural Resources, Division of Oil and Gas Resources Management

- Ohio Environmental Protection Agency
- Division of Air Pollution Control
- Division of Drinking and Ground Waters
- Division of Materials and Waste Management
- Division of Surface Water
- Pennsylvania Department of Environmental Protection, Office of Oil and Gas Management
- Railroad Commission of Texas, Oil and Gas Division
- Texas Commission on Environmental Quality
- Office of Air
- Office of Compliance and Enforcement
- Office of Legal Services
- Office of the Executive Director
- Wyoming Oil and Gas Conservation Commission
- Wyoming Department of Environmental Quality
- Air Quality Division
- Water Quality Division
- Solid and Hazardous Waste Division

Environmental organizations

- Dakota Resource Council
- Earthjustice

- Earthworks, Oil and Gas Accountability Project
- Pennsylvania Environmental Council

Public Health organizations

- American Lung Association
- Association of State and Territorial Health Officials
- Council of State and Territorial Epidemiologists

- National Association of County and City Health Officials
- Southwest Pennsylvania Environmental Health Project
- Trust for America's Health

Companies and industry organizations

- American Petroleum Institute
- Chesapeake Energy Corporation
- EOG Resources, Inc.

- Independent Petroleum Association of America
- North Dakota Petroleum Council

Source: GAO.

To identify additional requirements that apply to unconventional oil and gas development on federal lands, we reviewed laws, such as the National Environmental Policy Act (NEPA), as well as regulations and guidance promulgated by the Bureau of Land Management (BLM), Fish and Wildlife Service (FWS), Forest Service, and National Park Service. We also interviewed officials responsible for overseeing oil and gas development on federal lands, including officials in BLM headquarters and in field offices in the states we selected where there is a significant amount of oil and gas development on federal lands, including Colorado, North Dakota, and Wyoming; and in National Park Service, Forest

Service, and FWS headquarters. Oil and gas development may also be subject to tribal or local laws, but we did not include an analysis of these laws in the scope of our review.

To determine challenges that federal and state agencies face in regulating oil and gas development from unconventional reservoirs, we reviewed several reports conducted by environmental and public health organizations, industry, academic institutions, and government agencies that provided perspectives on federal and state regulations and associated challenges.[1] We also collected testimonial evidence, as described above, from knowledgeable federal and state officials, as well as industry, environmental, and public health organizations.

We conducted this performance audit from November 2011 to September 2012 in accordance with generally accepted government auditing standards. Those standards require that we plan and perform the audit to obtain sufficient, appropriate evidence to provide a reasonable basis for our findings and conclusions based on our audit objectives. We believe that the evidence obtained provides a reasonable basis for our findings and conclusions based on our audit objectives.

---

[1]For example, Ground Water Protection Council and ALL Consulting. "Modern Shale Gas Development in the United States: A Primer." Prepared for the Department of Energy and National Energy Technology Laboratory. April 2009. Natural Resources Defense Council. "Drilling Down: Protecting Western Communities from the Health and Environmental Effects of Oil and Gas Production." October 2007.

Appendix II: Key Requirements and Authorities under the Safe Drinking Water Act

The Safe Drinking Water Act (SDWA or the Act) was originally passed by Congress in 1974 to protect public health by ensuring a safe drinking water supply.[1] Under the act, EPA is authorized to set standards for certain naturally-occurring and man-made contaminants in public drinking water systems, among other things. Key aspects of SDWA for unconventional oil and gas development include provisions regarding underground injection and EPA's imminent and substantial endangerment authority.

Underground Injection Control Program

SDWA also regulates the placement of wastewater and other fluids underground through the Underground Injection Control (UIC) program.[2] This program provides safeguards to ensure that wastewater or any other fluid injected underground does not endanger underground sources of drinking water; these sources are defined by regulation as an aquifer or its portion:

1) (i) Which supplies any public water system; or

   (ii) Which contains a sufficient quantity of groundwater to supply a public water system; and (A) Currently supplies drinking water for human consumption; or (B) Contains fewer than 10,000 mg/l total dissolved solids; and

2) Which is not an exempted aquifer.[3]

Thus, the program is intended to protect not only those aquifers (or portions thereof) that are currently used for drinking water, but those that possess certain physical characteristics indicating they may be viable future drinking water sources.

EPA regulations establish criteria for exempting aquifers.[4] In particular, the regulations establish that the criterion that an aquifer "cannot now and

---

[1]Pub. L. No. 93-523, 88 Stat. 1660 (1974) (codified as amended at 42 U.S.C. §§ 300f–300j-26). Hereinafter, references to SDWA sections are as amended.

[2]SDWA §§ 1421-1426, 42 U.S.C. §§ 300h to 300h-5 (2012).

[3]40 C.F.R. § 146.3 (2012).

[4]40 C.F.R. § 146.4 (2012).

will not in the future serve as a source of drinking water" may be met by demonstrating that the aquifer is mineral, hydrocarbon or geothermal energy producing, or demonstrated by a permit applicant as having commercially producible minerals or hydrocarbons.[5] States or EPA typically initially identified exempt aquifers when UIC programs were established, and according to EPA, states may have added exempt aquifers since then. While EPA has the information from the initial applications, the agency does not have complete information for the additional exemptions, although under EPA regulations certain of these subsequent exemptions are considered program revisions and must be approved by EPA.[6] EPA is currently collecting information about the location of all exempted aquifers, and an official estimated that there are 1,000-2,000 such designations (including portions of aquifers).

There are six classes or categories of wells regulated through the UIC program.[7] Class II wells are for the management of fluids associated with oil and gas production, and they include wells used to dispose of oil and gas wastewater and those used to enhance oil and gas production.[8]

The EPA Administrator may approve by rule a state to have primary enforcement responsibility for the UIC program.[9] A state with an approved program assumes responsibility for implementing the program, including

---

[5]40 C.F.R. § 146.4 (2012).

[6]40 C.F.R. § 144.7(b)(3) ("For approved State programs exemption of aquifers identified (i) under § 146.04(b) shall be treated as a program revision under § 145.32"); § 146.4(b) (allowing exemption of aquifers that are producing or are economically producible for hydrocarbons); § 145.32 (establishing procedures for EPA approval of program revisions). According to EPA officials, nonsubstantial aquifer exemptions must be approved by the Regional Administrator, while substantial or major aquifer exemptions must be approved by the EPA Administrator. See also UIC Program Guidance 34.

[7]EPA regulations established well classes. 40 C.F.R. § 144.6 (2012).

[8]The other classes of UIC program wells are as follows: Class I wells are used for the disposal of hazardous and certain nonhazardous waste; Class III wells are used to inject fluids for mineral extraction; Class IV wells are used to dispose of hazardous or radioactive wastes, into or above an underground source of drinking water; Class VI wells are used for carbon sequestration. Class IV wells are currently banned. Class V wells are for any injection not covered by Classes I, II, III, IV, or VI.

[9]SDWA § 1422(b)(2), 42 U.S.C. § 300h-1(b)(2) (2012). See also SDWA §§ 1421(b)(1), 1422(b)(1), (3), 42 U.S.C. §§ 300h(b)(1), 300h-1(b)(1), (b)(3) (2012) (establishing requirements and responsibilities for states with primacy).

permitting, monitoring, and enforcement for UIC wells within the state. Generally, to be approved as the implementing authority (primacy), state programs must be at least as stringent as the federal program and show that their regulations contain effective minimum requirements for each of the well classes for which primacy is sought. Alternately, SDWA section 1425 provides that to obtain this authority over Class II wells only, a state with an existing oil and gas program may, instead of meeting and adopting the applicable federal regulations, demonstrate that its program is effective in preventing endangerment to underground sources of drinking water.[10] With respect to the six states in this review, Texas, North Dakota, Colorado, Wyoming, and Ohio have each been granted primacy for Class II wells under the alternative provisions (SDWA section 1425). EPA directly implements the entire UIC program in Pennsylvania.

Class II wells include saltwater (brine) disposal wells, enhanced recovery wells, and hydrocarbon storage wells. These wells are common, particularly in states with historical oil and gas activity. EPA officials estimate there are approximately 151,000 Class II UIC wells in operation in the United States; about 80 percent of these wells are for enhanced recovery, about 20 percent are for disposal, and there are approximately 100 wells for hydrocarbon storage. In Pennsylvania, the one state in our review in which EPA directly implements the Class II program, EPA Region 3 officials stated that there are five active Class II disposal wells. Recently, Region 3 issued permits for two Class II disposal wells in Pennsylvania, which were appealed. On appeal, the Environmental Appeals Board remanded the permits back to EPA for further consideration, finding that the Region failed to clearly articulate its regulatory obligations or compile a record sufficient to assure the public

---

[10]SDWA § 1425, 42 U.S.C. § 300h-4 (2012). As explained by EPA, under this alternative approval "instead of meeting the Federal Regulations (40 C.F.R. Parts 124, 144, and 145) and related Technical Criteria and Standards (40 C.F.R. Part 146), a State may demonstrate that its program meets the more general statutory requirements of Section 1421(b)(1)(A) through (D) and represents an effective program to prevent endangerment of underground sources of drinking water." See, e.g., 49 Fed. Reg. 13,040 (Apr. 2, 1984) (EPA approval of Colorado application). Thus, among other things, the state program must include adequate recordkeeping and reporting. The statute also provides that "[r]egulations of the Administrator under this section for State underground injection control programs may not prescribe requirements which interfere with or impede— (A) the underground injection of brine or other fluids which are brought to the surface in connection with oil or natural gas production or natural gas storage operations, or (B) any underground injection for the secondary or tertiary recovery of oil or natural gas." SDWA § 1421(b)(2), 42 U.S.C. § 300h(b)(2) (2012).

that the Region relied on accurate and appropriate data in satisfying its obligations to account for and consider all drinking water wells within the area of review of the injection wells.[11] The Environmental Appeals Board denied all other claims against EPA.[12] Under the remand, EPA may take further action consistent with the decision, which could include such actions as additions or revisions to the record and reconsideration of the permits. With respect to applications, according to Region 3 officials, until recently EPA did not receive many applications for new Class II brine disposal wells in Pennsylvania. EPA officials said that they have received five permit applications for such wells in the last 4 months and expect continued interest in the future.

Class II UIC Requirements

Under SDWA, UIC programs are to prohibit underground injection, other than into a well that is authorized by rule or permitted.[13] Class II UIC wells must meet requirements contained in either EPA regulations,[14] or relevant state regulations. Federal regulations for Class II wells include construction, operating, monitoring and testing, reporting, and closure requirements.[15] For example, one requirement of federal regulations is that all of the preexisting wells located in the area of review, and that were drilled into the same formation as the proposed injection well must be identified.[16] For such wells which are improperly sealed, completed, or abandoned, the operator must also submit a plan of actions necessary to prevent movement of fluid into underground sources of drinking water— known as "corrective actions," such as plugging, replugging, or operational pressure limits—which are considered in permit review.[17] Permits may be conditioned upon a compliance schedule for such corrective actions. According to EPA, in Pennsylvania many old wells

---

[11]See Environmental Appeals Board, Order, UIC Appeal No. 11-03 (June 28, 2012).

[12]Id.

[13]SDWA §§ 1421(b)(1)(A), 1422(c), 42 U.S.C. §§ 300h(b)(1)(A) (state programs), 300h-1(c) (EPA direct implementation programs); 40 C.F.R. § 144.11 (2012).

[14]See 40 C.F.R. pt. 144 (2012).

[15]40 C.F.R. §§ 146.21 – 146.24 (2012).

[16]40 C.F.R. §§ 146.24, 146.6 (2012).

[17]40 C.F.R. §§ 144.55(a), (b)(2)-(3); 146.7 (2012).

have had to be replugged in order to ensure they cannot present a potential pathway for migration.[18]

Regarding seismicity concerns, the federal regulations for Class II UIC wells require applicants for Class II UIC wells to identify faults if known or suspected in the area of review.[19] In addition there is a general requirement that a well must be sited to inject into a formation that is separated from any protected aquifer by a confining zone that is free of known open faults or fractures within the area of review.[20] In a permit process, EPA (in direct implementation states) or the state can require additional information (including geology) to ensure protection of underground sources of drinking water.[21] For example, Region 3 officials said the Region routinely determines whether there is the potential for fluid movement out of the injection zone via faults and fractures, as well as abandoned wells, by calculating a zone of endangering influence around the injection operation. Under the general standard, if a proposed or ongoing injection was, due to seismicity, believed to endanger underground sources of drinking water, EPA or the state could act, as the burden is on the applicant to show the injection well will not endanger such sources.[22] Officials said that if a seismic event occurs along a fault line that was not identified or known at the time of the UIC permit approval, EPA (in direct implementation states) or the state can go back to the well owner or operator and ask for additional information, which the owner or operator would be obligated to provide.

For additional information on the Class II UIC requirements applicable under EPA's program in Pennsylvania, see appendix IX.

---

[18]See Karen Johnson, Chief, Ground Water & Enforcement Branch, EPA Region 3, Marcellus Shale Educational Webinar, Feb. 18, 2010 (written Q&A).

[19]40 C.F.R. §§ 146.24, 146.24(a)(2) (2012).

[20]40 C.F.R. § 146.22(a) (2012).

[21]See 40 C.F.R. §§ 144.27(a), 144.51(h), 144.52(a)(9), 144.52(b)(1) (2012).

[22] 40 C.F.R. §§ 144.12(a), 144.1(b), (f) (2012).

Class II UIC Programs and Hydraulic Fracturing

Historically, UIC programs did not include hydraulic fracturing injections as among those subject to their requirements.[23] In 1994, in light of concerns that hydraulic fracturing of coalbed methane wells threatened drinking water, the Legal Environmental Assistance Foundation petitioned EPA to withdraw its approval of Alabama's Class II UIC program. EPA denied the petition, but on appeal, the United States Court of Appeals for the Eleventh Circuit held that the definition of underground injection included hydraulic fracturing and ordered EPA to reconsider the issue.[24] Subsequently, Alabama revised its program to include injection of hydraulic fracturing fluids,[25] and EPA approved it pursuant to SDWA section 1425 in 2000.[26] The Legal Environmental Assistance Foundation appealed the approval and, in 2001, the Eleventh Circuit partially remanded the approval, directing EPA to regulate hydraulic fracturing as Class II UIC wells rather than a Class II-like activity.[27] Alabama amended its regulations in 2001 and 2003.[28] EPA issued a determination in 2004 addressing the question on remand and found that the hydraulic fracturing portion of Alabama's UIC program relating to coalbed methane production, which was previously approved under the alternative effectiveness provision, complied with the requirements for Class II UIC wells.[29]

EPA initiated a study in 2000 to further examine the issue of fracturing in coalbed methane in areas of underground sources of drinking water.[30] EPA officials said the study showed diesel fuel was the primary risk. Subsequently, in 2003, EPA entered into a memorandum of agreement

---

[23]See, e.g., *Legal Envtl. Assistance Found. Inc. v. EPA*, 118 F.3d 1467, 1471 (11th Cir.1997).

[24]Id.

[25]Ala. Admin. Code r. 400-4-5-.04 (filed July 9, 1999; amended Nov. 9, 1999; repealed Apr. 11, 2000).

[26]65 Fed. Reg. 2889 (Jan. 19, 2000); 40 C.F.R. § 147.52 (2012). In May 2000, Alabama repealed the hydraulic fracturing regulations at 400-4-5-.04 and established regulations at 400 -3-8-.03. EPA's regulations reference the state's repealed regulations.

[27]*Legal Envtl. Assistance Found. Inc. v. EPA*, 276 F.3d 1253 (11th Cir. 2001).

[28]See Ala. Admin. Code r. 400 -3-8-.03.

[29]69 Fed. Reg. 42,341 (July 15, 2004) (referencing Alabama rule 400 -3-8-.03).

[30]EPA, Evaluation of Impacts to Underground Sources of Drinking Water by Hydraulic Fracturing of Coalbed Methane Reservoirs, EPA 816-R-04-003 (2004).

with three major fracturing service companies in which the companies voluntarily agreed to eliminate diesel fuel in hydraulic fracturing fluids injected into coalbed methane production wells in underground sources of drinking water.[31] According to EPA officials, the agreement is still in effect insofar as the agency has not received any termination notices.

EPA officials did not know of any permits issued by Alabama, or any other state, for hydraulic fracturing injections during this time frame. EPA also did not modify its direct implementation of Class II UIC programs to expressly include hydraulic fracturing.

On December 7, 2004, EPA's Assistant Administrator for Water responded to a congressional request for information on EPA's actions on this issue.[32] The letter summarizes EPA's study findings—that the potential threat to underground sources of drinking water posed by hydraulic fracturing of coalbed methane wells is low, but there is a potential threat through the use of diesel fuel as a constituent of fracturing fluids where coalbeds are colocated with an underground source of drinking water.[33]

The Eleventh Circuit court decision on the Alabama program generated significant controversy regarding whether hydraulic fracturing would be included in UIC programs nationwide. In this context, the Energy Policy Act of 2005[34] amended SDWA to include a provision exempting certain hydraulic fracturing injections from the UIC program. Specifically, the Energy Policy Act provided that "[t]he underground injection of fluids or propping agents (other than diesel fuels) pursuant to hydraulic fracturing operations related to oil, gas, or geothermal production activities" is excluded from the definition of "underground injection." Hence, injection of

---

[31]Memorandum of Agreement Between the United States Environmental Protection Agency and BJ Services Company, Halliburton Energy Services, Inc., and Schlumberger Technology Corporation, Elimination of Diesel Fuel in Hydraulic Fracturing Fluids Injected into Underground Sources of Drinking Water During Hydraulic Fracturing of Coalbed Methane Wells 4(a) (Dec. 12, 2003).

[32]151 Cong. Rec. S7277 (daily ed. June 23, 2005) (letter of Benjamin Grumbles, Assistant Administrator, Office of Water, EPA, to Senator Jeffords).

[33]Id.

[34]Pub. L. No. 109–58 § 322, 119 Stat. 594 (2005) (modifying SDWA § 1421(d)(1), 42 U.S.C. § 300h(d)(1) (2012)).

fluids other than diesel fuel in connection with hydraulic fracturing is not subject to federal UIC regulations, including both EPA direct implementation requirements and federal minimum requirements for state programs. The provision, however, did not exempt injection of diesel fuels in hydraulic fracturing from UIC programs.

EPA has prepared a draft guidance document to assist with permitting of hydraulic fracturing using diesel fuels under SDWA UIC Class II; a public comment period for this draft guidance closed in August 2012.[35] EPA explained that the guidance does not substitute for UIC Class II regulations, rather the guidance focuses on specific topics useful for tailoring Class II requirements to the unique attributes of hydraulic fracturing when diesel fuels are used.[36] EPA's draft guidance is applicable to any oil and gas wells using diesel in hydraulic fracturing (not just coalbed methane wells). The draft guidance provides recommendations related to permit applications, area of review (for other nearby wells), well construction, permit duration, and well closure. The guidance states that it does not address state UIC programs, although states may find it useful.

EPA officials told us that they recently identified wells for which publicly available data suggest diesel was used in hydraulic fracturing. EPA officials stated the agency also has some information on diesel use in hydraulic fracturing of shale formations from a 2011 congressional investigation. EPA officials said there are no EPA-issued permits authorizing diesel to be used in hydraulic fracturing, and they believe no applications for such permits have been submitted to EPA to date. EPA officials also said that they were not aware of any state UIC programs that had issued such permits.

Enforcement

Generally, EPA is authorized to enforce any applicable requirement of a federal or state UIC program as promulgated in 40 C.F.R. pt. 147, including Class II UIC programs approved under the alternative

---

[35]EPA, Permitting Guidance for Oil and Gas Hydraulic Fracturing Activities Using Diesel Fuels—Draft: Underground Injection Control Program Guidance #84 (May 2012 draft); see also 77 Fed. Reg. 27,451 (May 10, 2012) (announcing availability of the draft for public comment).

[36]77 Fed. Reg. at 27,452.

provision.[37] However, according to officials, EPA has not promulgated all of the states' modifications to UIC programs, and the federal regulations are out-of-date, hindering EPA's ability to directly enforce some state program provisions.[38]

EPA may issue administrative orders or, with the Department of Justice, initiate a civil action when a person violates any requirement of an applicable UIC program.[39] Where a state has primacy, EPA must first notify the state, and may act after 30 days if the state has not commenced an appropriate enforcement action.[40] SDWA also provides EPA with authority to access records, inspect facilities, and require provision of information. Specifically, EPA has authority, for the purpose of determining compliance, to enter any facility or property of any person subject to an applicable UIC program, including inspection of records, files, papers, processes, and controls.[41]

Under EPA's UIC program enforcement authorities, EPA has issued administrative compliance orders and administrative penalty orders relating to SDWA UIC Class II Wells. According to officials, most cases are administrative and handled at the Regional level. Officials said that there were more than 200 administrative orders related to the UIC program from 2004-2008 and that it is likely that a majority of these were related to Class II wells.

For example, EPA Region 3 signed a consent agreement in Venango County, Pennsylvania, where injections of produced water were made into abandoned wells not permitted under the UIC program.[42] In another

---

[37]SDWA requires that EPA approve state programs and revisions by regulation, in part 147 (rather than through an administrative process). SDWA § 1422(b)(2), (4), 42 U.S.C. § 300h-1(b)(2),(4) (2012).

[38]For federal regulations setting forth EPA-approved state programs, see 40 C.F.R. pt. 147 (2012).

[39]SDWA § 1423(a), 42 U.S.C. § 300h-2(a) (2012).

[40]SDWA § 1423(a)(1), 42 U.S.C. § 300h-2(a)(1) (2012).

[41]SDWA § 1445(b)(1), 42 U.S.C. § 300j-4(b)(1) (2012).

[42]Titusville Oil and Gas, Docket No. SDWA-03-20,11-0170 (July 19, 2011). EPA Region 3 officials noted that Pennsylvania Department of Environmental Protection also issued several orders at the site, and that the wells are being plugged.

case, Region 3 told us it has issued an administrative order against an operator for failure to conduct mechanical integrity tests. According to EPA, the order requires the operator to plug many of these wells, and to bring the wells they plan to continue to operate into compliance with their financial responsibility. Region 3 also took a penalty action against an operator for failure to report a mechanical integrity failure and continued operation after the failure.[43] According to officials, EPA was able to confirm during well rework that there was no fluid movement outside the well's casing and no endangerment to an aquifer.

Imminent and Substantial Endangerment Authorities

While SDWA generally does not directly regulate land use activities that may pose risk to drinking water supplies,[44] SDWA gives EPA authority to issue imminent and substantial endangerment orders or take other actions deemed necessary "upon receipt of information that a contaminant which is present in or is likely to enter a public water system or an underground source of drinking water...which may present an imminent and substantial endangerment to the health of persons, [where] appropriate State and local authorities have not acted to protect the health of such persons."[45] As noted above, the term "underground source of drinking water" includes not only active water supplies but also aquifers (or portions thereof) with certain physical characteristics.

EPA has used this imminent and substantial endangerment authority in several incidents where oil or gas wells have been alleged to contaminate drinking water. For example, EPA Region 8 has conducted long-term investigation and monitoring of groundwater contamination from an oilfield in Poplar, Montana, of a water supply serving Poplar, as well as the Fort Peck Indian Reservation. EPA determined that there are several plumes of produced water (brine) in the East Poplar aquifer, which supplies private and public drinking water wells. Several pathways of contamination have been identified, including unlined pits, spills, and a leaking plugged oil well.

---

[43]In the Matter of EXCO Resources (PA) LLC, Consent Agreement, EPA Docket No. SDWA-03-2012-0061 (Mar. 30, 2012).

[44]SDWA includes nonregulatory provisions addressing protection of drinking water supplies, such as providing incentives and assistance for states and public water systems to conduct water quality protection planning. See, e.g., SDWA § 1429, 42 U.S.C. § 300h-8 (2012).

[45]SDWA § 1431, 42 U.S.C § 300i(a) (2012).

EPA issued a SDWA imminent and substantial endangerment order in 2010 to three companies operating wells in the oilfield, each of which challenged the order in federal court. Following mediation, EPA and the parties entered an administrative order on consent in which the parties agreed to monitor the public drinking water supply for specified parameters and, if certain triggers are met or exceeded, to take actions to ensure the public water system meets water quality standards and pay reimbursement costs to the public water system.[46]

In another case, on December 7, 2010, EPA issued an administrative order to a well operator in Texas alleging methane contamination affecting private wells and directly related to its oil and gas production facilities.[47] EPA subsequently filed a complaint in U.S. District Court seeking injunctive relief to enforce the order's requirements and civil penalties for the operator's noncompliance with the order.[48] A few days later, the operator filed a petition for review of the order with the Fifth Circuit Court of Appeals. The operator's position was that the order is not a final agency action and that EPA has the burden of proving its claim in the district court enforcement action, and its enforcement would violate due process.[49] On March 29, 2012, EPA withdrew its administrative order, and the parties moved for voluntary dismissal of both cases.[50] In a letter to EPA, the operator agreed to conduct sampling of 20 private water wells for 1 year.[51]

---

[46]EPA, Fort Peck East Poplar Oil Field Safe Drinking Water Act Emergency Administrative Order on Consent, Docket No. SDWA-08-20 12-0019 (Mar. 26, 2012); see also EPA, Emergency Administrative Order, Docket No. SDWA-08-2011-0006 to Murphy Exploration & Production Co.-USA, Pioneer Natural Resources USA, Inc., and SGH Enterprises, Inc. (Dec. 16, 2010).

[47]In the Matter of Range Resources Corporation, Administrative Order, EPA Docket No. SDWA-06-2011-1208 (Dec. 7, 2010).

[48]United States v. Range Production Co., No. 3:11-cv-00116-F (N.D. Tex. Jan. 18, 2011).

[49]Brief, Range Resources Corporation, et al. v. EPA, No. 11-60040 (5th Cir. Mar. 22, 2011).

[50]See Joint Stipulation of Dismissal Without Prejudice, United States v. Range Production Co., No. 3:11-CV-00116-F (N.D. Texas Mar. 30, 2012); see also Range Resources Corporation, et al. v. EPA, No. 11-60040 (5th Cir. Mar. 30, 2012).

[51]Letter, Bracewell & Giuliani to EPA Office of Enforcement & Compliance Assurance, Mar. 30, 2012.

Appendix III: Key Requirements and Authorities under the Clean Water Act

Under the Clean Water Act (CWA),[1] EPA regulates discharges of pollutants to waters of the United States; for the purpose of this document, we generally refer to such waters, including jurisdictional rivers, streams, wetlands, and other waters, as surface waters.[2] Discharges may include wastewater, including produced water, and stormwater. In addition, together with the U.S. Army Corps of Engineers, EPA regulates discharge of dredged or fill material into these waters.[3]

Under CWA section 311 and the Oil Pollution Act,[4] EPA regulations establish, in relevant part, requirements for the prevention of, preparedness for, and response to oil discharges at certain facilities, including among others oil drilling and production facilities.[5] These requirements may include Facility Response Plans and Spill Prevention, Control, and Countermeasure (SPCC) Plans. EPA also has certain response and enforcement authorities relevant to these requirements.

This review focuses on EPA regulatory activities under these programs relevant to unconventional oil and gas development activities.

---

[1]The Federal Water Pollution Control Act Amendments of 1972, Pub. L. No. 92-500, § 2, 86 Stat. 816 (amending the Act of June 30, 1948, ch. 758, 62 Stat. 1155) (codified as further amended at 33 U.S.C. ch. 26, §§ 1251-1387 (2012) and commonly referred to as the Clean Water Act). Hereinafter, references are to CWA sections as amended.

[2]For the purpose of this document, when we use the term "surface waters" in relation to federal regulation, we refer to waters of the United States, including jurisdictional rivers, streams, wetlands, and other waters. State definitions of the term "surface waters" may differ. EPA officials noted that some surface waters may not be jurisdictional for certain CWA provisions.

[3]The U.S. Army Corps of Engineers administers the day-to-day program, including issuance of permits and enforcement, among other things. EPA has the opportunity to review and comment on individual permit applications, can enforce CWA section 404 provisions, and has authority to veto Corps permit decisions as to the discharge of dredged or fill material at defined sites, among other things. See CWA §§ 404, 404(c), 33 U.S.C. §§ 1344, 1344(c) (2012); http://water.epa.gov/lawsregs/guidance/cwa/dredgdis/404c_index.cfm

[4]CWA § 311, 33 U.S.C. § 1321 (2012); Oil Pollution Act of 1990, Pub. L. No. 101-380, 104 Stat. 484 (classified as amended at 40 U.S.C. ch. 40, §§ 2701 – 2761 (2012) and amending sections of CWA). See also Exec. Order 12,777, 56 Fed. Reg. 54,757 (1991).

[5]40 C.F.R. pt. 112 (2012).

National Pollutant Discharge Elimination System Program

CWA is the primary federal law designed to restore and maintain the chemical, physical, and biological integrity of the nation's waters. Among other things, EPA and delegated states[6] administer CWA's National Pollutant Discharge Elimination System (NPDES) program, which limits the types and amounts of pollutants that facilities such as industrial and municipal wastewater treatment plants may discharge into the nation's surface waters.[7] Facilities such as municipal wastewater treatment plants and industrial sites, including oil and gas well sites, need a permit if they have a point source discharge to surface waters. Other than stormwater runoff as discussed below, discharges of pollutants from an oil or gas well site to surface water require an NPDES permit. According to EPA, wastewater associated with shale gas extraction can include total dissolved solids, fracturing fluid additives, metals, and naturally occurring radioactive materials, and may be disposed by transport to publicly-owned or other wastewater treatment plants, particularly in some locations where brine disposal wells are unavailable.[8] According to EPA, produced water from coalbed methane gas extraction can include high salinity and pollutants such as chloride, sodium, sulfate, bicarbonate, fluoride, iron, barium, magnesium, ammonia, and arsenic,[9] and some produced water is discharged to surface water in certain geographical areas.[10]

---

[6]Of the states in our review, EPA is the NPDES permitting authority for the oil and gas industry in Texas, and also administers the pretreatment program in Colorado, Pennsylvania, and Wyoming.

[7]CWA also features a system of water quality standards consisting of designated uses and water quality criteria, expressed as constituent concentrations, levels, or narrative statements, representing a quality of water that supports the use. Water quality standards play a critical role in the act's framework. For example, if technology based limitations are insufficient to meet water quality standards, then more stringent water quality based limitations are to be added to discharge permits. EPA is currently updating chloride water quality criteria with a draft criteria document expected in 2012; a more stringent chloride criteria could eventually affect permit limits for any facility discharging oil and gas wastewater. See www.epa.gov/hydraulicfracture/.

[8]76 Fed. Reg. 66,286, 66,296 (Oct. 26, 2011).

[9]See www.epa.gov/hydraulicfracture/ and 76 Fed. Reg. at 66,293-97. EPA conducts annual reviews of existing effluent guidelines and pretreatment standards and biennially publishes a plan identifying the industrial categories selected for new or revised rules.

[10]See http://water.epa.gov/scitech/wastetech/guide/cbm_index.cfm

EPA and delegated states issue discharge permits that set conditions in accordance with applicable technology-based effluent limitations guidelines that EPA has established for various industrial categories, and may also include water-quality based effluent limitations.[11] When EPA issues effluent limitations guidelines for an industrial category, it may include both limitations for direct dischargers (point sources that introduce pollutants directly into waters of the United States) and pretreatment standards applicable to indirect dischargers (facilities that discharge into publicly-owned wastewater treatment plants).[12]

Existing Effluent Limitations Guidelines for Oil and Gas Extraction

EPA has developed effluent limitations guidelines for several subcategories of the oil and gas extraction industry.[13] The guidelines generally apply to facilities engaged in the production, field exploration, drilling, well completion, and well treatment in the oil and gas extraction industry.[14] The guidelines applicable to the wells in the scope of this review—essentially, oil and gas wells located on land and drilling unconventional reservoirs—include those for the onshore subcategory, agricultural and wildlife water use subcategory, and stripper wells.[15] The guidelines for these subcategories were finalized in 1979.[16]

---

[11]Water-quality based effluent limitations are imposed when technology-based limitations are insufficient for receiving waters to meet water quality standards.

[12]See, e.g., 76 Fed. Reg. at 66,288.

[13]40 C.F.R. pt. 435 (2012).

[14]See, e.g., 40 C.F.R. § 435.30 (2012).

[15]40 C.F.R. §§ 435.30-.32 (onshore), 435.50-.52 (agricultural and wildlife water use), 435.60-.61 (stripper wells) (2012). EPA also has developed effluent limitation guidelines for the subcategories offshore wells and coastal wells; because this report is focused on onshore wells, we do not discuss the guidelines for the offshore subcategory. See 40 C.F.R. §§ 435.10-.15 (offshore), 435.40-.47 (coastal) (2012). EPA's original 1979 rule included coastal wells in the subcategory for onshore wells. Following a court order, EPA in 1982 suspended the applicability of the guidelines for the onshore category to coastal wells. 44 Fed. Reg. 22,069 (Apr. 13, 1979), 47 Fed. Reg. 31,554 (July 21, 1982). See also *American Petroleum Institute v. EPA*, 661 F.2d. 340 (5th Cir. 1981).

[16]In 1976, EPA issued interim final regulations establishing effluent limitation guidelines for the oil and gas extraction category and, in 1979, EPA replaced these with final guidelines. 44 Fed. Reg. 22,069 (Apr. 13, 1979); see also 47 Fed. Reg. 31,554 (July 21, 1982).

For the onshore and agricultural and wildlife water use subcategories,
EPA established effluent limitations guidelines for direct dischargers. EPA
did not establish guidelines for stripper wells, explaining that
unacceptable economic impacts would occur from use of the then-
evaluated technologies, and that the agency could revisit this decision at
a later date.[17] EPA officials we spoke with said that they are not aware of
any reconsideration of this decision, and that this is not an issue on the
current regulatory agenda. EPA also did not establish pretreatment
requirements for either onshore or stripper well subcategories.[18]

Existing effluent limitations guidelines do not apply to wastewater
discharges from coalbed methane extraction.[19] As EPA subsequently
explained, because there was no significant coalbed methane production
in 1979, the oil and gas extraction rulemakings did not consider coalbed
methane extraction in any of the supporting analyses or records.[20] EPA
officials also told us that the coalbed methane process is fundamentally
different than traditional oil and gas exploration because of the volume of
water that must be removed from the coalbed before production can
begin, which they see as a significant distinction for potentially applicable
technology. As will be discussed later in this appendix, in October 2011,
EPA announced its intention to develop effluent limitations guidelines and
standards for wastewater discharges from the coalbed methane industry.

When an oil and gas well proposing to discharge pollutants to a surface
water is not covered by the existing guidelines, effluent limitations
included in the permit are determined on a case-by-case basis by the

---

[17]41 Fed. Reg. 44,942, 44,946-47 (Oct. 13, 1976). EPA has also noted that the stripper
subcategory may be used to exclude a well from other subcategories. In effect, the
existence of the subcategory authorizes a permit writer to set case-specific permit
limitations for a well that falls within the stripper subcategory, as it has no specific effluent
limitations, rather than use those for the onshore subcategory. 44 Fed. Reg. at 22,073.

[18]By definition, such requirements would be inapplicable to the agricultural and wildlife use
subcategory.

[19]76 Fed. Reg. 66,286, 66,293 (Oct. 26, 2011).

[20]EPA, Technical Support Document for the 2006 Effluent Guidelines Program Plan, EPA-
821R-06-018, 6-1 (2006).

relevant permitting authority, using best professional judgment[21] and any applicable state rules or guidance. EPA officials were not aware of any other unconventional oil and gas extraction processes, besides coalbed methane extraction, that are not covered by the existing effluent limitations guidelines.

Table 8 summarizes the coverage and key requirements of the existing guidelines.

Table 8: Summary of Effluent Limitations Guidelines for Wastewater Discharges from Selected Subcategories of Oil and Gas Wells Located on Land

Subcategory of oil or gas well	Covered by existing effluent limitations guideline	Permit requirement
Wells included in an existing segment subcategory		
Onshore wells that do not fit into any of the other subcategories	Yes – direct dischargers No – indirect dischargers (no pretreatment requirements established)	No discharge
Agricultural and Wildlife Water Use subcategory	Yes	Discharge of produced water allowed where conditions met, and subject to max daily limit for oil and grease
Stripper wells	Subcategory created, but no effluent limitations or pretreatment requirements established	Case-by-case
Wells that are not included in any existing segment subcategory		
Coalbed methane extraction wells	No	Case-by-case

Source: GAO analysis of federal regulations.

Note: EPA proposed pretreatment standards for oil and grease for new sources in the four subcategories in 1979, but did not issue them. Compare 41 Fed. Reg. 44,949, 44,952 (Oct. 13, 1976), 44 Fed. Reg. 22,069 (Apr. 13, 1979) and 76 Fed. Reg. 66,286, 66,295 (Oct. 26, 2011).

Onshore Subcategory

The effluent limitations guideline for the Oil and Gas Extraction point source category, onshore subcategory establishes the best practicable control technology currently available as

[21]EPA has noted that permit writers are to develop technology-based limits on a case-by-case basis using their best professional judgment, and considering the same statutory factors EPA would use in promulgating a national categorical effluent limitation guideline. See 40 C.F.R. §§ 122.44(a)(1),125.3(d) (2012). EPA, Technical Support Document for the 2006 Effluent Guidelines Program Plan, EPA-821R-06-018, 6-3 (2006).

there shall be no discharge of waste water pollutants into navigable waters from any source associated with production, field exploration, drilling, well completion, or well treatment (i.e., produced water, drilling muds, drill cuttings, and produced sand).[22]

Because an NPDES permit is only required where a facility discharges or proposes to discharge a pollutant, and as the technology-based requirement of "no discharge" must be applied in the permit, facilities subject to a "no discharge" limit are not required to apply for such permits.

According to the 1976 *Federal Register* Notice of the Proposed Rule, technologies for managing produced water to achieve no discharge to surface waters were expected to include evaporation ponds, or underground injection, either for enhanced recovery of oil or gas in the producing formation or for disposal to a deep formation.[23] Further, EPA indicated that drilling muds, drill cuttings, well treatment wastes, and produced sands would be disposed by land disposal so as not to reach navigable waterways.

Agricultural and Wildlife Water Use Subcategory

The effluent limitations guideline for the Oil and Gas Extraction point source category also established a subcategory for Agricultural and Wildlife Water Use to cover a geographical subset of operations in which produced water is of good enough quality to be used for wildlife or livestock watering or other agricultural uses and that the produced water is actually put to such use during periods of discharge. This subcategory guideline is only applicable to facilities located west of the 98th meridian, which extends from approximately the eastern border of North Dakota south through central Texas. EPA explained in the preamble to this rule that "[i]t is intended as a relatively restrictive subcategorization based on the unique factors of prior usage in the Region, arid conditions and the existence of low salinity, potable water."[24]

For this subcategory, the guideline establishes the best practicable control technology currently available as

---

[22]40 C.F.R. § 435.32 (2012) (emphasis added).

[23]41 Fed. Reg.44,942, 44,946 (Oct. 13, 1976).

[24]44 Fed. Reg. 22,069, 22,072 (Apr. 13, 1979).

"no discharge of waste pollutants into navigable waters from any source (other than produced water) associated with production, field exploration, drilling, well completion, or well treatment (i.e., drilling muds, drill cuttings, and produced sands)," and for produced water discharges a daily maximum limitation of 35 milligrams per liter of oil and grease.[25]

At oil and gas well sites meeting the conditions of location, produced water quality, and use of produced water for wildlife or livestock watering or agricultural use, the produced water may be discharged to waters of the United States. In terms of water quality, the produced water must be "good enough" for this use,[26] and must not exceed the daily maximum for oil and grease. States generally issue these permits, and are responsible for determining whether the water is of appropriate water quality for the beneficial use.[27] EPA is responsible for oversight and has not issued guidance on this topic.

EPA has not revised the guildeines, such as to add limitations for additional pollutants, to define "good enough" water quality, or to establish potentially more stringent guidelines. EPA officials stated that it has not done so because in certain locations the produced water from oil and gas development is high quality, and because treatment would cost more than injection, thus discouraging the beneficial use of this water.

With respect to the subcategories of oil and gas wells covered by the effluent limitations guidelines, discharges are authorized only for oil and gas wells under the Agricultural and Wildlife Water Use and Stripper well subcategories. These well sites that discharge wastewater to surface waters must, as noted above, obtain a NPDES permit from the permitting authority (state, tribe, or EPA). The permit is to incorporate the applicable effluent limitations guideline, if one exists, and include effluent monitoring and reporting requirements. Officials also stated that individual permits may contain limits for pollutants other than oil and grease.

---

[25]40 C.F.R. §§ 435.50-52 (2012).

[26]40 C.F.R. § 435.51(c) (2012).

[27]EPA issues permits in a small number of states that do not have responsibility for the NPDES program, and on tribal lands.

According to EPA, 349 discharge permits in the Agricultural and Wildlife Water Use subcategory have been issued. Most of these permitted discharges are located in Wyoming, Montana, and Colorado.

Anticipated Rulemaking to Develop Effluent Limitations Guidelines for Oil and Gas Extraction from Coalbed Methane Formations

On October 26, 2011, EPA announced in its Final 2010 Effluent Limitations Program Plan that the agency will develop effluent limitations guidelines and standards for wastewater discharges from the coalbed methane extraction industry.[28]

With respect to coalbed methane extraction, as noted above, there is no existing effluent limitations guideline applicable to associated wastewaters. Coalbed methane operations discharging wastewaters to surface waters must nonetheless obtain a NPDES permit, but in the absence of a federal effluent limitations guideline, the permitting authority determines the permit limits based on best professional judgment, as well as any applicable state rules or guidelines. EPA had identified the industry for consideration in prior years, and initiated work leading to a detailed study beginning in 2007.[29] EPA's 2010 coalbed methane study found that states are primarily issuing individual permits, but they are also issuing some general permits and watershed permits covering one or more wells through a streamlined process.[30] According to EPA officials, eastern states have generally based effluent limitations in permits on the coal mining effluent limitations guideline, although that guideline does not have limitations for total dissolved solids or chlorides that are key components of produced water. In the six states reviewed, EPA identified 861 coalbed methane discharge permits.[31] According to EPA officials, most coalbed methane wastewater discharges have NPDES permits.

Following the study, EPA concluded that some of the waters discharged to surface waters have high total dissolved solids, and that there are readily available technologies to treat this produced water and decided to

[28]76 Fed. Reg. 66,286 (Oct. 26, 2011).

[29]76 Fed. Reg. at 66,293. See also
http://water.epa.gov/scitech/wastetech/guide/cbm_index.cfm

[30]EPA, Coalbed Methane Extraction: Detailed Study Report, EPA-820-R-10-022 (2010).

[31]Id. at Appendix A. See also EPA, Technical Support Document for the 2006 Effluent Guidelines Program Plan, EPA-821R-06-018, 6-5 (2006) (listing numbers of coalbed methane discharge permits derived from state permit databases).

initiate rulemaking.[32] EPA is in the preproposal stage of rulemaking for the coalbed methane effluent guidelines and standards.[33] EPA's website indicates the projected date for publication of the proposed rule is June 2013.

Generally Applicable Pretreatment Standards and POTW Obligations

Facilities discharging industrial wastewater to publicly-owned treatment works (POTW) treatment plants are subject to general pretreatment requirements. In addition, the POTW receiving such industrial wastewaters also has responsibilities related to its own permit and to receiving these wastewaters.

General standards

EPA has issued general pretreatment requirements applicable to all existing and new indirect dischargers of pollutants (other than of purely domestic, or sanitary, sewage) to a POTW, including any dischargers of wastewaters associated with oil and gas wells.[34] Notably, such discharges are subject to a general requirement that the pollutants do not cause pass through or interference with the POTW.[35] For a discharge to cause pass through, it must contribute to violation of the POTW's NPDES permit; to cause interference, it must contribute to the noncompliance of its sewage sludge use or disposal.[36]

Other standard provisions for indirect discharges involve a prohibition on corrosive discharges.[37] According to EPA officials, in produced water,

---

[32]76 Fed. Reg. at 66,294.

[33]See http://yosemite.epa.gov/opei/RuleGate.nsf/byRIN/2040-AF35?opendocument#1, RIN 2040-AF35, docket no. EPA-HQ-OW-2011-0334.

[34]See generally 40 C.F.R. pt. 403 (2012).

[35]40 C.F.R. § 403.5(a)(1) (2012).

[36]Pass through means a discharge that exits the POTW into waters of the United States in quantities or concentrations which, alone or in conjunction with a discharge or discharges from other sources, is a cause of a violation of any requirement of the POTW's NPDES permit. 40 C.F.R. § 403.3(p) (2012). Interference is a discharge which, alone or in conjunction with a discharge or discharges from other sources, both (1) inhibits or disrupts the POTW, its treatment processes or operations, or its sludge processes, use or disposal; and (2) therefore is a cause of a violation of any requirement of the POTW's NPDES permit or prevents sewage sludge use or disposal in compliance with relevant laws. 40 C.F.R. § 403.3(k) (2012).

[37]40 C.F.R. § 403.5(b)(2) (2012).

concerns for corrosivity would be related to high chlorides and sulfides which could adversely affect pipes and gaskets in the POTW.[38]

EPA has stated that NPDES permits for POTWs typically do not contain effluent limits for some of the pollutants of concern from shale gas wastewater, and that some of these pollutants may be harmful to aquatic life.[39] Specifically, if a POTW did not include information in its NPDES permit application indicating that the POTW would receive oil and gas wastewater, or did not otherwise adequately characterize the incoming wastewater as including certain pollutants of concern, the permit may not include limits for these pollutants, as permits generally only contain limits for those pollutants reasonably expected to be present in the wastewater.[40]

Regarding pass through, in which an indirect industrial discharger contributes to violation of the receiving POTW's NPDES permit, Region 3 officials said that POTW operators had not indicated that NPDES violations were caused by oil and gas wastewaters received at the plant, with the following exception. In 2011, EPA issued an administrative order for compliance and request for information to a POTW in New Castle, Pennsylvania, in relation to permit effluent limit violations.[41] The POTW experienced violations of its suspended solids limits spanning over a year, and attributed the violations to salty wastewater from natural gas production it was receiving. The order required the POTW to take several actions including to cease accepting oil and gas exploration and

---

[38]See also 74 Fed. Reg. 58,784, 58,803 (Nov. 13, 2009) (in the context of SPCC regulations, stating "Information reviewed by the Agency and presented in the public docket (EPA–HQ–OPA–2007–0584–0015) showed corrosion as a common cause of oil and produced water discharges at onshore oil production facilities. The higher salt content of produced water fluids as compared to crude oil may lead to the increased corrosion rate of metallic components of the produced water storage system.").

[39]Memorandum from EPA Office of Wastewater Management to EPA Regions with answers to frequently asked questions about wastewater issues resulting from shale gas extraction (2012), available at http://cfpub.epa.gov/npdes/hydrofracturing.cfm

[40]NPDES permits held by POTWs do typically require monitoring for whole effluent toxicity, intended to measure whether the effluent is harmful to aquatic life. Some POTWs in Pennsylvania have previously accepted produced water; EPA Region 3 officials were unsure if data are available regarding whether these POTWs had problems with their whole effluent toxicity tests.

[41]In the Matter of New Castle Sanitation Authority, Findings of Violation Order for Compliance and Request for Information, EPA. Docket No. CWA-03-2011-0272 DN.

production wastewater until completing an evaluation and sampling, and
to eliminate and prevent recurrence of the violations.

POTW Obligations

Generally, local governments operating POTWs are responsible for
ensuring that indirect dischargers comply with any applicable national
pretreatment standards.[42] Certain POTWs are required to develop
pretreatment programs, which set out a facility's approach to developing,
issuing, and enforcing pretreatment requirements on any indirect
dischargers to the particular plant.[43] EPA or states may be responsible for
ensuring these POTWs meet their obligations and for approving the
POTW's pretreatment plans.

According to EPA, regardless of pass through or interference, POTWs
should not accept indirect discharges of produced water if the
wastewaters have different characteristics than those for which the
POTW was originally permitted, without providing adequate notice to the
permitting authority.[44] If a POTW accepts oil and gas wastewater with
characteristics that were not considered at the time of the permit
issuance, then the permit may not adequately protect the receiving water
from potential violations of water quality standards. In other words, a
POTW may meet its permit limits, yet still contribute to a violation of water
quality standards, if the permit does not reflect consideration of all the
pollutants actually present, and their concentrations, in the incoming
wastewater and in the discharge. According to Region 3 officials, EPA
has conducted several investigations of whether discharges from POTWs
accepting oil and gas wastewater have prevented receiving waters from
meeting water quality standards. Region 3 officials stated that a major
impediment to this evaluation was that the NPDES permits reviewed did
not have effluent limits or monitoring requirements for the pollutants of
concern. EPA also stated that it has data from a 2009 Pennsylvania
Department of Environmental Protection violation report documenting a

[42]See EPA, Introduction to the National Pretreatment Program, EPA-833-B-11-001,3-3
(2011).

[43]C.F.R. § 403.8(a) (2012); see also http://cfpub.epa.gov/npdes/faqs.cfm?program_id=3

[44]Memorandum from EPA Office of Wastewater Management to EPA Regions with
Answers To Frequently Asked Questions About Wastewater Issues Resulting From Shale
Gas Extraction, Attachment at 9 (Mar. 16, 2011), available at
http://cfpub.epa.gov/npdes/hydrofracturing.cfm

fishkill attributed to a spill of diluted produced water in Hopewell
Township, PA.[45]

In March 2011, EPA's Office of Water issued to the Regions a set of
questions and answers that provide state and federal permitting
authorities in the Marcellus shale region with guidance on permitting
treatment and disposal of wastewater from shale gas extraction.[46] The
guidance states that POTWs must provide adequate notice to the
permitting authority (EPA or the authorized state) of any new introduction
of pollutants into the POTW from an indirect discharger, if the discharger
would be subject to NPDES permit requirements if it were discharging
directly to a surface water, among other things.[47] EPA officials indicated
that if a POTW is accepting types of wastewater that were not on its
original application, EPA could require a modification of the POTW's
NPDES permit, or object to a NPDES renewal that did not address these
wastewaters and the facility's ability to treat them. POTWs may also
initiate inclusion of these wastewaters in their permits or permit renewals.
For example, EPA Region 3 officials stated that four POTW operators in
Pennsylvania in the NPDES renewal process have indicated the intent to
continue accepting oil and gas wastewater. In addition, in cases with pass
through or interference, EPA could require a POTW to develop a
pretreatment program.

EPA's website indicates the agency plans to supplement the existing
Office of Water questions and answers document with additional
guidance directed to permitting authorities, pretreatment control
authorities and POTWs, to provide assistance on how to permit POTWs
and other centralized wastewater treatment facilities by clarifying existing
CWA authorities and obligations.[48] Specifically, EPA plans to issue two
guidance documents, one for permit writers and another for POTWs.

---

[45]76 Fed. Reg. 66,286, 66,297 (Oct. 26, 2011).

[46]Memorandum from EPA Office of Wastewater Management to EPA Regions with
Answers To Frequently Asked Questions About Wastewater Issues Resulting From Shale
Gas Extraction (Mar. 16, 2011), available at
http://cfpub.epa.gov/npdes/hydrofracturing.cfm

[47]Id. at 9; see also 40 C.F.R. § 122.42(b)(1) (2012).

[48]http://www.epa.gov/hydraulicfracture/#swdischarges

Anticipated Rulemaking to Develop Pretreatment Standards for Gas Extraction from Shale Formations

With respect to shale gas extraction, the effluent limitations guideline for the onshore subcategory in effect since 1979 has prohibited direct discharges of associated wastewaters; however, EPA has not established pretreatment standards for indirect discharges of such wastewaters. EPA requested and received comments on whether to initiate a rulemaking for the industry in recent years.[49]

In 2011, EPA announced it will initiate a rulemaking to develop such pretreatment standards. EPA reviewed existing data, but did not conduct a study to develop data as it had for coalbed methane. EPA found that pollutants in wastewaters associated with shale gas extraction are not treated by the technologies typically used at POTWs or many centralized treatment facilities.[50] Further, EPA stated that resulting discharges have the potential to affect drinking water supplies and aquatic life. On this basis, EPA concluded that pretreatment standards are appropriate and decided to initiate a rulemaking.[51] EPA intends to conduct a survey, among other things, to collect information on management of produced water to support the rulemaking.[52] Finally, EPA noted that if it obtains information indicating that POTWs are already adequately treating shale gas wastewater, the agency could adjust the rulemaking plans accordingly.[53] For example, the state of Pennsylvania requested that operators of Marcellus shale gas wells stop delivering produced water to POTWs, potentially avoiding the issue. EPA officials stated that other states may nonetheless have a need to utilize POTWs to address these wastewaters and hence could benefit from pretreatment standards.

EPA is in the preproposal stage of this rulemaking, and EPA's website indicates the projected date for publication of the proposed rule is 2014.

---

[49]See, e.g., 76 Fed. Reg. at 66,292, 66,295, 74 Fed. Reg. 68,599 (Dec. 28, 2009).

[50]76 Fed. Reg. at 66,295-96. According to EPA, POTWs typically have permits that do not contain limits for the pollutants of concern in shale gas wastewater; the secondary treatment requirements do not address such pollutants, and is it uncommon for these permits to contain water quality based limitations for such pollutants. Id. at 66,297. Thus, such wastewaters likely pass through the POTWs receiving such wastewaters and the POTWs may not monitor for these pollutants in their effluent.

[51]Id. at 66,297.

[52]Id.

[53]Id. at 66,298.

NPDES for Stormwater Discharges

In 1987, the Water Quality Act amended CWA to establish a specific program for regulating stormwater discharges of pollutants to waters of the United States.[54] Among other things, the amendments clarified EPA authority to require an NPDES permit for discharges of stormwater from several categories, including in relevant part those associated with industrial activity and construction activity.[55] EPA subsequently issued regulations that address stormwater discharges from several source categories, including certain industrial activities and construction activities.[56]

Stormwater from Industrial Activities

Generally, industrial sites obtain coverage for stormater through a general permit, such as the multisector general permit or construction general permit.[57] To do so, the facility operator submits a notice of intent, and agrees to meet general permit conditions. For example, conditions for the construction general permit include applicable erosion and sediment control, site stabilization, and pollution prevention requirements.[58]

In providing EPA authority to regulate stormwater discharges, the Water Quality Act also prohibited EPA from requiring a NPDES permit for discharges of stormwater from:

oil and gas exploration, production, processing, or treatment operations or transmission facilities composed entirely of flows which are from conveyances or systems of conveyances (including but not limited to pipes, conduits, ditches, and channels) used for collecting and conveying precipitation runoff and which are not contaminated by contact with, or do not come into contact with, any overburden,[59] raw material,

---

[54]See generally Water Quality Act of 1987, Pub. L. No. 100-4 § 405, 101 Stat. 7, 69-71 (adding CWA § 402(p), codified at 33 U.S.C. § 1342(p) (2012)).

[55]Id., 55 Fed. Reg. 47,990, 47,992 (Nov. 16, 1990) (Phase 1 rule).

[56]55 Fed. Reg. at 47,990.

[57]See Multisector General Permit at App. C.

[58]See EPA, 2012 Construction General Permit Fact Sheet at 6-7, 36.

[59]Subsequently, EPA added to its regulations a definition of overburden: "any material of any nature, consolidated or unconsolidated, that overlies a mineral deposit, excluding topsoil or similar naturally-occurring surface materials that are not disturbed by mining operations." 40 C.F.R. § 122.26(b)(10) (2012).

intermediate products, finished product, byproduct, or waste products located on the site of such operations.[60]

Interpreting the provision exempting oil and gas facilities, EPA issued regulations requiring permits for *contaminated* stormwater from oil and gas facilities.[61] To determine whether a discharge of stormwater from an oil or gas facility is contaminated, EPA regulations establish that if a facility has had a stormwater discharge that resulted in a discharge exceeding an EPA reportable quantity requiring notification under the Comprehensive Environmental Response, Compensation, and Liability Act (CERCLA) or section 311 of CWA, or which contributes to violation of a water quality standard, the permit requirement is triggered for that facility.[62]

Regarding stormwater at oil and gas well sites, officials said it is unlikely there is a permit requirement because it is rare that stormwater would come into contact with raw materials. Nonetheless, if a facility anticipates having a stormwater discharge that includes a reportable quantity of oil or may result in a violation of water quality standards, then the facility would be obligated to apply for a NPDES permit. In applying for the permit, however, the facility has to agree not to discharge pollutants in a reportable quantity and not to discharge pollutants so as to cause a water quality violation. Given this, it is unclear whether facilities would apply for such a permit after they have had a release of a reportable quantity or contributing to a water quality violation. Furthermore, according to officials, EPA relies upon operators self-identifying based on reportable quantities or water quality violations.

Despite these factors, EPA reviewed available data for the five states in which EPA administers the NPDES program,[63] including Texas, and

---

[60]Water Quality Act of 1987, § 401, 101 Stat. 65-66 (codified at 33 U.S.C. § 1342(l)(2) (2012)).

[61]55 Fed. Reg. at 48,029, 40 C.F.R. §§ 122.26(b)(14)(iii), (c)(1)(iii) (2012).

[62]Id. The reporting requirements triggering a permit are listed in 40 C.F.R. §§ 117.21, 302.6, 110.6 (2012). Note that the generally applicable industrial stormwater regulations also provide a conditional exclusion for discharges composed entirely of stormwater if there is no exposure of industrial materials and activities to rain, snow, snowmelt and runoff, and where additional conditions are satisfied. 40 C.F.R. § 122.26(g) (2012).

[63]The states are Idaho, Massachusetts, Texas, Oklahoma, and Alaska.

identified some stormwater general permit notifications for facilities that could be well sites.[64]

Stormwater from Construction Activity

EPA regulations require permits for stormwater discharges from construction activities including clearing, grading, and excavating that result in land disturbance. Beginning in 1990, EPA began regulating stormwater discharges from construction sites disturbing more than 5 acres of land under its Phase I rule. Under Phase II rules issued in 1999, EPA regulated stormwater discharges from construction sites disturbing between 1 and 5 acres of land, with initial permit applications due in 2003.

With respect to oil and gas well sites, under the statutory provisions and EPA's Phase 1 stormwater regulations, discharges of stormwater from construction activity would have required a permit only for sites disturbing more than 5 acres and where the stormwater is contaminated by contact with, or comes into contact with, any overburden, raw material, intermediate products, finished product, byproduct, or waste products located on the site of such operations.[65] According to EPA officials, the agency believed few oil and gas sites met these conditions. They further explained that when EPA conducted the Phase II rulemaking for the smaller 1 to 5 acre sites, the agency assumed incorrectly that oil and gas well sites would be smaller than 1 acre and thus did not include oil and gas well sites in their economic analysis of the rule. After the rule's issuance as it became aware that such sites would fall under the rule, and in light of industry objections over the lack of economic analysis, EPA delayed Phase II implementation at oil and gas well sites until 2006.[66]

---

[64]Specifically, EPA officials identified approximately 34 notifications from facilities with Standard Industrial Classification codes that could indicate that the facility is a well site for example, codes 1381 (drilling oil and gas) and 1389 (hydraulic fracturing services).

[65]CWA §§ 402(*l*)(2), 502(24), 33 U.S.C. §§ 1342(*l*)(2), 1362(24); 55 Fed. Reg. 47,990, 48,065 (Nov. 16, 1990) (amending 40 C.F.R. § 122.26; see § 122.26(a)(1)(ii), (b)(14)(x)).

[66]In 2005, EPA revised its regulation to extend the deadline for permits for contaminated stormwater discharges associated with small construction activity—generally including clearing, grading and excavating that results in land disturbance more than 1 acre but less than 5 acres—at oil and gas sites to June 12, 2006. 70 Fed. Reg. 11,560, 11,563 (Mar. 9, 2005) (amending 40 C.F.R. § 122.26(e)(8)). Other industry sites requiring permit coverage for small construction activity were required to comply by March 2003.
http://cfpub.epa.gov/npdes/stormwater/cgp.cfm

Before implementation of Phase II regulations at oil and gas well sites began, the Energy Policy Act of 2005 was enacted. The Energy Policy Act of 2005 amended CWA to specifically define the activities included in the oil and gas stormwater exemption. Where the law already exempted from NPDES permit requirements discharges of stormwater from "oil and gas exploration, production, processing, or treatment operations or transmission facilities,"[67] the Energy Policy Act of 2005 added a definition of this term as "all field activities or operations associated with exploration, production, processing, or treatment operations, or transmission facilities, including activities necessary to prepare a site for drilling and for the movement and placement of drilling equipment, whether or not such field activities or operations may be considered to be construction activities."[68]

In response to these amendments, in 2006, EPA revised a key provision of the regulations concerning oil and gas stormwater discharges.[69] The revision provided that discharges of sediment from oil or gas facility construction activities and contributing to a water quality standard violation would not trigger a permit requirement. This revision was vacated and remanded by the Ninth Circuit in 2008.[70] EPA has not subsequently revised the regulations applicable to stormwater discharges from oil and gas facilities; the pre-2006 regulations remain in effect as to this industry.[71] EPA officials said the agency intends to revise its regulations to address the court's vacatur in an upcoming stormwater rulemaking,[72] with the proposal expected in 2013.

According to EPA officials, during construction, oil and gas well sites would have no permit requirement because of the statutory exemption.

---

[67]Water Quality Act of 1987 § 401 (adding 33 U.S.C. § 1342(l)(2)).

[68] Energy Policy Act of 2005, Pub. L. No. 109–58, § 323, 119 Stat. 594, 694 (adding 33 U.S.C. § 1362(24) (2012)).

[69]71 Fed. Reg. 33,628 (June 12, 2006) (amending 40 C.F.R. § 122.26(a)(2)(ii) (2006)). As stated in EPA's fact sheet for the 2006 revisions, permits would be required only in "very limited instances."

[70]*Natural Resources Defense Council v. EPA*, 526 F.3d 591 (9th Cir. 2008).

[71]Id. See also
http://cfpub.epa.gov/npdes/regresult.cfm?program_id=6&type=1&sort=name&view=all

[72]Compare 40 C.F.R. § 122.26 (2012) with 40 C.F.R. § 122.26 (2005).

NPDES Enforcement

For violations of the law, or applicable regulations or permits, EPA has authority to issue administrative orders requiring compliance, impose administrative penalties, as well as to bring suit and, in conjunction with the Department of Justice, to impose civil penalties.[73] Among other things, EPA can take such actions if a well operator violates the CWA prohibition on unauthorized discharges of pollutants to surface waters.[74] EPA also has information-gathering and access authority relative to point source owners and operators, which could include certain oil and gas well site operations.[75] For example, EPA has authority to inspect facilities where an effluent source is located.[76]

As an example of enforcing the prohibition of unauthorized discharges, in 2011, EPA Region 6 assessed an administrative civil penalty against a company managing an oil production facility in Oklahoma for discharging brine and produced water to a nearby stream.[77] In another case, EPA entered a consent agreement with an oil production company in Colorado for unauthorized discharges of produced water from a multiwell site due to a failed gas eliminator valve in a produced water transportation pipeline.[78] The produced water travelled overland for 333 feet, then entered a stream tributary to an interstate river. The company agreed to pay a civil penalty and to conduct a macroinvertebrate study for the affected watershed.

Imminent and Substantial Endangerment Authorities

CWA section 504 provides EPA an imminent and substantial endangerment authority, authorizing EPA to bring suit to restrain a person to stop the discharge of pollutants causing or contributing to pollution, or to

---

[73]CWA § 309, 33 U.S.C. § 1319 (2012).

[74]Id., CWA § 301(a), 33 U.S.C. § 1311(a) (2012). As explained above, however, discharges of pollutants via stormwater from a well site generally does not require a permit, and thus such discharges without a permit would not be considered unauthorized, except in limited circumstances.

[75]CWA § 308(a), 33 U.S.C. § 1318(a) (2012).

[76]CWA § 308(a)(B)(i), 33 U.S.C. § 1318(a)(B)(i) (2012).

[77]American Petroleum & Environmental Consultants, Inc., Cease and Desist Administrative Order, EPA Docket No. CWA-06-2012-1760 (Dec. 12, 2011).

[78]BP America Production Company, Combined Complaint and Consent Agreement, EPA Docket No. CWA-08-2012-0014 (May 15, 2012).

take such other action as may be necessary, upon receipt of evidence that a pollution source or combination of sources is presenting an imminent and substantial endangerment to the health of persons or to the livelihood of persons.[79] Unlike the analogous provisions of several other major environmental laws, however, CWA section 504 does not expressly mention administrative orders.

Oil and Hazardous Substances Spill Prevention, Reporting, and Response

Spill Prevention and Response Plans

EPA's Oil Pollution Prevention regulations, promulgated and amended pursuant to CWA and the Oil Pollution Act, impose spill prevention and response planning requirements on oil and gas well sites that meet thresholds.[80] Specifically, the Spill Prevention, Control, and Countermeasure (SPCC) Rule applies to sites with underground and/or aboveground storage tanks above certain thresholds and where oil could be discharged into or upon navigable waters.[81] Onshore oil and gas production facilities, among others, generally are subject to the rule if they (1) have an aggregate oil storage capacity of greater than 1,320 gallons in aboveground oil storage containers or a total oil storage capacity greater than 42,000 gallons in completely buried storage tanks and (2) could reasonably be expected, due to their location, to discharge harmful

---

[79]CWA § 504, 33 U.S.C. § 1364 (2012).

[80]40 C.F.R. pt. 112 (2012); see also 67 Fed. Reg. 47,042 (July 17, 2002).

[81]Id.

quantities of oil into or upon U.S. navigable waters or adjoining shorelines.[82]

The SPCC rule, as amended, requires each owner or operator of a regulated facility to prepare and implement a plan that describes how the facility is designed, operated, and maintained to prevent oil discharges into or upon U.S. navigable waters and adjoining shorelines. The plan must also include measures to control, contain, clean up, and mitigate the effects of these discharges.

EPA regulations specify requirements for SPCC plans for onshore oil drilling and oil production facilities.[83] Onshore drilling facilities must meet the general requirements for such plans, as well as meet specific discharge prevention and containment procedures: (1) position or locate mobile drilling or workover equipment so as to prevent a discharge; (2) provide catchment basins or diversion structures to intercept and contain discharges of fuel, crude oil, or oily drilling fluids; and (3) install a blowout prevention (BOP) assembly and well control system before drilling below any casing string or during workover operations.[84] Oil production facilities are exempt from the SPCC security provisions.[85]

EPA has amended the SPCC regulations from time to time. In 2008, EPA amended the regulations to streamline certain requirements, to be effective in January 2010.[86] These amendments included several provisions easing requirements for oil production sites, such as excluding them from loading/unloading rack requirements, providing alternative

---

[82]40 C.F.R. § 112.2 (2012) (For purposes of the SPCC rule, "[t]he term 'navigable waters' of the United States means 'navigable waters' as defined in section 502(7) of the FWPCA, and includes: (1) all navigable waters of the United States, as defined in judicial decisions prior to the passage of the 1972 amendments of the Federal Water Pollution Control Act, (FWPCA) (Pub. L. No. 92-500) also known as CWA and tributaries of such waters as; (2) interstate waters; (3) intrastate lakes, rivers, and streams that are utilized by interstate travelers for recreational or other purposes; and (4) intrastate lakes, rivers, and streams from which fish or shellfish are taken and sold in interstate commerce.").

[83]40 C.F.R. § 112.10 (2012).

[84]Id.

[85]40 C.F.R. § 112.7(g) (2012). See also U.S. Chemical Safety and Hazard Investigation Board, Investigative Study Final Report: Public Safety at Oil and Gas Storage Facilities, Report No. 2011-H-1 at 8 (September 2011).

[86]73 Fed. Reg. 74,236 (Dec. 5, 2008).

qualified facility eligibility criteria (for streamlined compliance with self-certification of SPCC plans), and exempting certain produced water containers.[87] In November 2009, EPA revised the amendments, eliminating these three provisions before they became effective.[88] EPA explained that, in consideration of relevant facts and public comments, there was either no basis for the exclusion or the exclusion will not effectively protect the environment from discharges. For example, with respect to the alternative qualified facility criteria provision, EPA stated that the agency

> reviewed the spill data for the oil production sector contained in its study of the exploration and production sector...While these data do not characterize the extent of environmental damage caused by oil discharges from small oil production facilities, they demonstrate that the volume of oil discharged from onshore oil production facilities [is] increasing, and the number of oil discharges on a yearly basis has remained the same, despite a decline in crude oil production. In addition, oil production facilities are often unattended, and typically located in remote areas, which potentially increases the risk of environmental damage from an oil discharge of oil.[89]

Various development activities at oil and gas well sites involve storage of oil that may trigger the SPCC regulations to impose these requirements. During initial exploration and drilling, the capacity of the fuel tank of the drill rig is the primary way the SPCC rule could be triggered, and EPA officials said that almost all drill rigs exceed the threshold capacity. During well completion and workover, where hydraulic fracturing is conducted, EPA officials said the capacity of the fuel tank in the turbines and pumps being used for fracturing typically exceed the threshold. As to wells in the production phase, they said there would generally be no SPCC requirement at dry gas wells, because they would not be storing condensate on-site. For wet gas and oil production, the size of the condensate or oil tanks on the site would be the key to whether SPCC is triggered.

---

[87]74 Fed. Reg. 58,784, 58,799-803 (Nov. 13, 2009).

[88]Id. at 58,799-803, 58,809-811.

[89]Id. at 58,802. See also Considerations for the Regulation of Onshore Oil Exploration and Production Facilities Under the Spill Prevention, Control, and Countermeasure Regulation (40 C.F.R. part 112)) (available at www.regulations.gov document EPA–HQ–OPA–2007–0584–0015).

EPA has developed guidance related to SPCC applicability and compliance for oil production, drilling, and workovers.[90] According to officials, EPA is currently developing a "frequently asked questions" document about the SPCC program and hydraulic fracturing. This document is being developed in response to an influx of questions about how the SPCC rule applies to gas well sites, particularly from companies active in the Marcellus shale. According to EPA officials, while the SPCC is focused on oil, wet gas wells involve condensates, some of which have traditionally been deemed liquid hydrocarbons and included in the program. In particular, questions have arisen over the lightest condensates (C2 and C4 hydrocarbons), which are usually in gaseous form at standard temperatures and pressures and hence are not included in the SPCC program, whereas storage of heavier condensates, such as C6+ hydrocarbons, has been included (as liquids) in the SPCC program.

EPA directly administers the SPCC program.[91] EPA's regulations do not require facilities to report to the agency that they are subject to the SPCC rule and, as of 2008, EPA did not know the universe of SPCC-regulated facilities, but the agency was considering developing some data.[92] EPA officials stated that they have significant data but not complete data because of the lack of a registration or submittal requirement. To ensure that facility owners and operators are meeting SPCC requirements, EPA personnel inspect selected regulated facilities to determine their compliance with the regulations. For some facilities, the SPCC compliance date was in November 2011.[93] EPA is working to develop a national database of sites inspected under the SPCC rule. Officials said that the SPCC program's database includes 120 inspections at oil and gas production facilities for fiscal year 2011, of which 105 had some form of noncompliance, which varies in significance from paperwork inconsistencies to more serious violations (though EPA officials were unable to specifically quantify the number of more serious violations).

---

[90]See www.epa.gov/osweroei/content/spcc/spcc_up.htm, http://www.epa.gov/Region8/opa/wkshop.html (Cross Reference Matrix For Drilling And Workover Facilities).

[91]The Clean Water Act does not provide EPA with the authority to authorize states to implement the program in its place.

[92]GAO-08-482.

[93]See 40 C.F.R. § 112.3(a)(1) (2012); see generally 40 C.F.R. § 112.3(a)(1)-(3) (2012) (establishing deadlines for SPCC plans).

According to EPA headquarters officials, EPA generally selects facilities for inspection based on spill reports EPA receives through the National Response Center.

The Oil Pollution Prevention Regulation also requires an owner or operator of nontransportation onshore facilities that could, because of location, reasonably be expected to cause substantial harm to the environment by discharging oil into or on the navigable waters or shorelines, to submit to the appropriate EPA Regional office a facility response plan.[94] The regulation specifies criteria to be used in determining whether a facility could reasonably be expected to cause substantial harm and hence triggers such requirement,[95] and it also provides that the EPA Administrator may at any time, on determination considering additional factors, require a facility to submit a facility response plan.[96] A facility owner or operator also may maintain certification that it could not, because of location, reasonably be expected to cause substantial harm by discharging oil into or onto navigable waters or shorelines.[97] Relevant to oil well sites, the initial criteria for requiring a facility response plan is that the facility has total oil storage of 1 million gallons or more. Where such facilities meet at least one of four other criteria—such as lacking secondary containment, or located at distances that could injure fish and wildlife—then a facility response plan is required.[98] The plan is to provide, in essence, an emergency response action plan for the worst-case discharge and other relevant information.[99] According to EPA officials, onshore oil well sites would typically not go over the threshold criteria triggering the requirement for a facility response plan. Officials said there may be a small number of sites where very large or centralized operations with a number of wells connected to central piping and/or storage might trigger a facility response plan.

---

[94]40 C.F.R. § 112.20 (2012).

[95]Id. at (f)(1), 40 C.F.R. Pt. 112, App. C, Att. C-1.

[96]40 C.F.R. §§ 112.20(b)(1), (c), (f) (2012).

[97]Id. at (e).

[98]Id. at (f)(1), App. C, Att. C-1 (2012). For new facilities, the facility response plan is to be submitted before startup, for facilities commencing operation after Aug. 30, 1994; time frames are different for facilities required to submit a plan due to changes in operation, among other things. See 40 C.F.R. § 112.20(a)(2)(ii) (2012).

[99]Id. at (h).

Spill Prohibition and Reporting

CWA established the policy of the United States that there should be no discharges of oil or hazardous substances into or upon U.S. navigable waters or onto adjoining shorelines, among other resources, and generally prohibited such discharges.[100] Relevant provisions require reporting of certain discharges of oil or a hazardous substance to these waters.[101]

EPA has issued regulations designating those hazardous substances that present an imminent and substantial danger to the public health or welfare when discharged to U.S. navigable waters or onto adjoining shorelines in any quantity.[102]

EPA also has determined, in regulations, the quantities of oil and other hazardous substances of which the discharge to U.S. navigable waters or onto adjoining shorelines may be harmful to the public health or welfare or the environment.[103] CWA in conjunction with these regulations require facilities to report to the National Response Center certain unpermitted releases of oil or hazardous substances to surface waters.[104] The National Response Center subsequently sends reports to EPA Regions and headquarters. With respect to oil, discharges of oil must be reported if they "(c)ause a film or sheen upon or discoloration of the surface of the water or adjoining shorelines or cause a sludge or emulsion to be deposited beneath the surface of the water or upon adjoining shorelines," or if they violate applicable water quality standards.[105] With respect to

---

[100]CWA § 311(b)(1), (3), 33 U.S.C. § 1321(b)(1), (3) (2012). The scope of jurisdiction for the section 311 oil spill program is broader than that for the NPDES program. The section 311 oil spill program and the NPDES program both have jurisdiction over navigable waters of the United States; Section 311 also provides jurisdiction over spills of oil or hazardous substances into or on adjoining shorelines or that may affect natural resources of the United States, among others.

[101]Id. at (b)(3)-(5).

[102]40 C.F.R. § 116.4 (2012), see also CWA § 311(b)(2)(A), 33 U.S.C. § 1321(b)(2)(A) (2012).

[103]CWA § 311(b)(2)(A), (4), 33 U.S.C. § 1321(b)(2)(A), (4) (2012), 40 C.F.R. pt. 110, pt. 116-17 (2012).

[104]CWA § 311(b), 33 U.S.C. § 1321(b) (2012), 40 C.F.R. 110.6 (2012). See also 33 C.F.R. § 153.203 (2012).

[105]40 C.F.R. §§ 110.6, 110.3, 110.1 (2012),
http://www.epa.gov/osweroe1/content/reporting/faq_subs.htm#pelist

hazardous substances, EPA has determined threshold quantities—those which may be harmful to the public health or welfare or the environment— known as reportable quantities.[106]

Spill Response Authority

EPA, as well as other relevant federal agencies, has various response authorities to ensure effective and immediate removal of a discharge, and mitigation or prevention of a substantial threat of a discharge, of oil or a hazardous substance to U.S. navigable waters or onto adjoining shorelines.[107] The National Oil and Hazardous Substances Pollution Contingency Plan, issued by EPA by regulation, provides a system to respond to discharges and to contain, disperse, and remove oil and hazardous substances, among other things.[108] For example, according to EPA Region 5, in conjunction with the state of Ohio, the Region has responded to several incidents in which orphan wells were found to be leaking or discharging crude oil into waterways.

Under CWA section 311, as required to carry out its purposes including spill prevention and response, EPA also has authority to require the owner or operator of a facility subject to the Oil Pollution Prevention Regulation, among other provisions, to establish and maintain such records; make such reports; install, use, and maintain such monitoring equipment and methods; provide such other information deemed necessary; and for entry and inspection of such facilities.[109]

Enforcement of SPCC and Spill Prohibition and Reporting Requirements

For violations of the law, or applicable regulations or permits, EPA has authority to issue administrative orders requiring compliance, impose administrative penalties, as well as to bring suit, in conjunction with the Department of Justice, to impose civil penalties.[110] Section 311 also gives EPA the authority to access records and inspect facilities, and the ability

---

[106]40 C.F.R. §§ 117.3; § 116.4 at table 116.4A; see also 40 C.F.R. § 302.4 at table 302.4 (2012).

[107]CWA § 311(c), 33 U.S.C. § 1321(c) (2012).

[108]Id., 40 C.F.R. pt. 300 (2012).

[109]CWA § 311(m)(2), 33 U.S.C. § 1321(m)(2) (2012).

[110]Id. at (b)(6)-(7), (c), (e)(1)(B) (2012).

to require provision of information, with respect to persons and facilities
subject to section 311, including SPCC program requirements.[111]

For example, in Region 8, EPA participated in an effort with the U.S. Fish
and Wildlife Service (FWS), states, and tribes, after FWS expressed
concerns about migratory birds landing on open pits that contained oil
and water, which killed or harmed the birds.[112] This effort involved aerial
surveys to observe pits. Where apparent problems were identified,
relevant federal or state agencies were notified and were to give oil and
gas operators an opportunity to correct problems. Ground inspections
were then conducted where deemed warranted and, if problematic
conditions were found, further follow-up action was taken by EPA or the
relevant state or other federal agency. As a result of this effort, 99 sites
with violations of SPCC requirements were identified.[113] EPA's report
stated that "[n]on-compliance with SPCC requirements was more
pervasive than anticipated. Although the SPCC program has been the
focus of outreach and compliance assistance nationally for more than 25
years, there remains a strong need to communicate its requirements,
inspect regulated facilities, and conduct appropriate technical assistance
or enforcement to ensure improved compliance."[114] The report states that,
for most SPCC violations, EPA issued a notice of violation and that many
notice of violation recipients came into compliance without escalation to
formal enforcement, but that some enforcement actions were taken.[115]
Region 8 reported identifying 22 sites with documented SPCC violations
as a result of subsequent efforts in 2004-2005.[116] Information on the
nature or resolution of these violations was not readily available.

---

[111]Id. at (m)(2) (2012).

[112]EPA Region 8, Oil and Gas Environmental Assessment Effort 1996 – 2002, v (2003).

[113]Id. at 7.

[114]Id. at 8.

[115]Id. at 8.

[116]EPA Region 8, Summary Report, Oil and Gas Environmental Assessment Activities in
Wyoming during 2004-2006.

Imminent and Substantial Endangerment Authority for Spills

CWA section 311 provides EPA authority to address certain releases of oil or hazardous substances to U.S. navigable waters and adjoining shorelines. Specifically, on determination that "there may be an imminent and substantial threat to the public health or welfare of the United States, including fish, shellfish, and wildlife, public and private property, shorelines, beaches, habitat, and other living and nonliving natural resources under the jurisdiction or control of the United States, because of an actual or threatened discharge of oil or a hazardous substance from a vessel or facility" in violation of the prohibition against discharges of oil or hazardous substances to U.S. navigable waters and adjoining shorelines, EPA may bring suit, or may, after notice to the affected state, take any other action under this section, including issuing administrative orders, that may be necessary to protect the public health and welfare.[117]

---

[117]CWA § 311(e)(1), 33 U.S.C. § 1321(e)(1) (2012). EPA also may, through the Department of Justice, file a civil action to secure any relief from any person, including the facility owner or operator, as may be necessary to abate such endangerment.

Appendix IV: Key Requirements and Authorities under the Clean Air Act

The Clean Air Act (CAA or the Act) is the primary law with the purpose to protect and enhance the nation's air quality.[1] Under CAA, EPA regulates two primary types of air pollutants: criteria pollutants and hazardous air pollutants (HAPs). EPA sets, and periodically may revise, National Ambient Air Quality Standards for six criteria pollutants—carbon monoxide, sulfur dioxide, lead, nitrogen dioxide, particulate matter, and ozone.[2] States then develop state implementation plans (SIP) and seek EPA approval; if a SIP is not acceptable, EPA may assume primary responsibility for implementing and enforcing CAA in that state.[3] In addition, EPA retains CAA implementation and enforcement oversight authority in states with approved SIPs. The SIPs generally establish how the state will attain air quality standards, through regulation, permits, policies, and other means. SIPs also must demonstrate that the state program satisfies certain minimum federal statutory and regulatory requirements. EPA also establishes various federal air emission regulations addressing criteria pollutants or HAPs, and which generally fall into two categories of emission sources: stationary sources and mobile sources. Stationary sources of air pollution are generally any building, structure, facility, or installation that may emit any air pollutant.[4] With respect to oil and gas production as a stationary source, EPA has identified various components of oil and gas production that may be relevant to emissions:

> Production components may include, but are not limited to, wells and related casing head, tubing head and "Christmas tree" piping, as well as pumps, compressors, heater treaters, separators, storage vessels, pneumatic devices and dehydrators. Production operations also include the well drilling, completion and workover processes and includes all the portable non-self-propelled apparatus associated with those operations.[5]

---

[1]Clean Air Act Amendments of 1970, Pub. L. No. 91-604, 84 Stat. 1676 (1970) (codified as amended at 42 U.S.C. §§ 7401-7671q (2011)((commonly referred to as the Clean Air Act). Hereinafter, references to CAA are as amended.

[2]See CAA § 109, 42 U.S.C. § 7409 (2012).

[3]CAA § 110, 42 U.S.C. § 7410 (2012).

[4]CAA §§ 302(z), 111(a)(3), 112(a)(3), 42 U.S.C. §§ 7602(z), 7411(a)(3), 7412(a)(3) (2012), 40 C.F.R. §§ 52.21(b)(5), 60.2, 63.3 (2012).

[5]76 Fed. Reg. 52,738, 52,744 (Aug. 23, 2011). EPA's description of oil and gas production also includes certain facilities separate from the well pad, such as pipelines transporting oil and gas to refineries or processing plants, which are outside the scope of GAO's review.

In addition, EPA officials have noted that tanks, ponds, and pits are sources of emissions that may be present at well sites. Others have also identified condensate storage tanks and flaring as significant emission sources often associated with gas wells.[6] The key criteria pollutant of concern for oil and gas production is VOCs, as an ozone precursor, and the primary HAP released by the oil and gas production industry are BTEX (benzene, toluene, ethylbenzene, and xylenes) and n-hexane.[7]

To address stationary sources under CAA, EPA is required to promulgate industry-specific emissions standards such as National Emission Standards for Hazardous Air Pollutants (NESHAP) and New Source Performance Standards (NSPS) for source categories that EPA has listed under the Act.[8] CAA also provides for review of new and modified major sources of emissions under the Prevention of Significant Deterioration and Nonattainment New Source Review programs, typically implemented by states.[9] CAA and EPA regulations require operating permits, known as Title V permits, for certain stationary sources, and establish minimum requirements for state operating permitting programs.[10] Each of these key programs is described below as it may apply to oil and gas well sites.

Mobile sources associated with oil and gas production may include trucks bringing fuel, water, and supplies to the well site; construction vehicles; and truck-mounted pumps and engines. That is, oil and gas wells may be served by a variety of road and nonroad vehicles and engines. EPA regulates emissions from an array of mobile sources by imposing emission limits on such vehicles and engines; these generally applicable regulations are not specific to the oil and gas industry and are not discussed here.

Finally, the Act includes provisions addressing accidental releases of dangerous pollutants to the air.[11] Oil and gas wells are unlikely to trigger the planning aspects of these provisions, according to EPA; however, the

---

[6]NRDC, Drilling Down: Protecting Western Communities from the Health and Environmental Effects of Oil and Gas Production 11 (2007).

[7]64 Fed. Reg. 32,610 (June 17, 1999).

[8]CAA §§ 111(b)(, (f), 112(c)(2), (d), 42 U.S.C. §§ 7411(b), (f), 7412(c)(2), (d) (2012).

[9]CAA §§ 165(a), 171-193, 160-169B, 42 U.S.C. §§ 7475(a), 7501-7515, 7470-7492 (2012).

[10]CAA § 502, 42 U.S.C. § 7661a (2012), 40 C.F.R. §§ 70.3, 71.3 (2012).

[11]CAA § 112(r), 42 U.S.C. § 7412(r) (2012).

well sites are subject to the general duty clause, a self-implementing provision of CAA under which operators are responsible for identifying hazards associated with accidental releases and designing and maintaining a safe facility, taking such steps as are necessary to prevent releases.

Table 9 summarizes the applicability of key Clean Air Act programs to emission points at oil and gas well sites. These provisions will be discussed in greater detail in this appendix.

Table 9: Summary of CAA Programs That May Apply to Emissions from Oil and Gas Well Sites

CAA program	Specific rule for oil and gas production sector, applicable to well sites?	Likely applicability to emissions at oil and gas well sites
National Emission Standards for Hazardous Air Pollutants (NESHAP)	Yes, addressing glycol dehydrators and storage vessels with potential for flash emissions at major sources and only triethylene glycol dehydrators at area sources[a]	Data not available regarding the extent to which well sites comprise major or area sources subject to the NESHAP, but the NESHAPs may apply at few well sites.
New Source Performance Standards (NSPS)	Yes, addressing gas well completions, pneumatic controllers, and some storage vessels	Gas wells that are fractured are subject to flaring and phased-in green completion requirements, focused on VOC reductions, under April 2012 rule. (Oil wells that are fractured are not subject to NSPS.) Data not available on how many pneumatic controllers are at wells (in total, 13,500 subject to rule). Data not available on how many storage vessels over threshold of 6 tons per year of VOC are at wells. In total, EPA estimates that 304 new source storage vessels will be subject to the rule annually and that most of these storage vessels are expected to be at wells.
New Source Review (NSR)	No (generally applicable)	Unknown; most likely is nonattainment NSR in severe areas, which have lowest threshold potential to emit. Programs are generally implemented by states and features vary.
Title V Operating Permits	No. The oil and gas NESHAP and NSPS specifically exempt area or nonmajor sources subject to these rules from Title V permitting requirements if they are not otherwise required by law to obtain such permits.	Unknown.
Accidental Releases: Risk Management Program	No (generally applicable)	Unlikely; regulated substances in oil and gas that meet the definition of "naturally occurring hydrocarbon mixture" are not counted toward threshold quantities of regulated substances; extent to which other regulated substances are present in threshold quantities at well sites not known.
Accidental Releases: General Duty Clause	No (generally applicable)	Applies; no threshold; EPA has used provision to conduct inspections and require control of leaks.
Greenhouse Gas Reporting	Yes	Applies with threshold at the basin-level, EPA estimates that certain emissions from approximately 467,000 onshore wells will be covered.

Source: GAO.

[a]With respect to dehydrators, the major source standards apply to large glycol dehydrators—those with an actual annual average natural gas flowrate equal to or greater than 85 thousand standard cubic meters per day and actual annual average benzene emissions equal to or greater than 0.90 Mg per year. The area source standards apply to triethylene glycol dehydration units with natural gas flow rate at or above 3 million standard cubic feet per day and benzene emission at or above 1 ton per year.

National Emission Standards for Hazardous Air Pollutants

Hazardous Air Pollutants

The 1990 CAA amendments significantly expanded the hazardous air pollutants program; they identified 189 specific HAPs to be regulated, required EPA to list categories of sources to be regulated, and established implementation timelines.[12] The list of HAPs includes several potentially found in oil and gas well emissions.[13] In addition to these listed HAPs, EPA and others have identified hydrogen sulfide, which is found in oil and gas well emissions but is not a listed HAP, as hazardous and toxic to humans.[14] EPA has the authority to add to the HAPs list pollutants which may present, through inhalation or other routes of exposure, a threat of adverse human health effects or adverse environmental effects, but not including releases subject to EPA's regulation under section 112(r)—namely, the accidental release and risk management regulations.[15] The prevention of accidental releases regulation includes accidental releases of hydrogen sulfide.[16] In a 1993 report to Congress,

---

[12]Clean Air Act Amendments of 1990, Pub. L. No. 101-549, Title III, § 301, 104 Stat. 2399, 2531; Joint resolution to make a technical correction in Public Law 101-549, Pub. L. No. 102-187, 105 Stat. 1285 (1991). Of the 189 HAPs, two have since been delisted, so the list now includes 187 HAPs.

[13]The listed HAPs are also known as air toxics.

[14]In other regulatory contexts, hydrogen sulfide is considered hazardous; for example, it is a hazardous substance under CERCLA. 40 C.F.R. § 302.4 table 302.4, (2012). See also EPA, *Report to Congress on Hydrogen Sulfide Emissions Associated with the Extraction of Oil and Gas*, EPA-453-R-93/045 (1993) (stating that hydrogen sulfide "is toxic and care should be exercised in its presence;" EPA officials noted that the purpose of this report was not to examine whether or not hydrogen sulfide should included in the HAPs list.). Some have argued that hydrogen sulfide, of which oil and gas well emissions may be a significant source, should be a HAP. See NRDC, Drilling Down, at 13. While it was on the list initially enacted in the 1990 amendments, by a joint resolution, Congress corrected the inadvertent addition of hydrogen sulfide to the list. See page 2 of GAO, *Clean Air Act: EPA Should Improve the Management of Its Air Toxics Program*, GAO-06-669 (Washington, D.C.: Jun. 23, 2006). Cf. Earthworks and Oil & Gas Accountability Project, The Oil and Gas Industry's Exclusions and Exemptions to Major Environmental Statutes 14 (2007).

[15]CAA § 112(b)(2), 42 U.S.C. § 7412(b)(2) (2012).

[16]40 C.F.R. § 68.130 (2012).

EPA found that the limited data available did not evidence a significant threat to human health or the environment from "routine" emissions of hydrogen sulfide from oil and gas wells.[17]

CAA provides a process to petition EPA to modify the HAPs list. On March 30, 2009, the Sierra Club and 21 other environmental and public health organizations and individuals petitioned EPA to list hydrogen sulfide as a HAP under section 112(b).[18] The petitioners asserted that low-level hydrogen sulfide emissions not addressed by the accidental release provisions in section 112(r) are harmful to human health.[19] EPA officials told us they are considering the petition but have no specific timeline for acting upon it.

NESHAPs Overview and Statutory Provisions Restricting Aggregation of Oil and Gas Production Sources

EPA is required to promulgate and periodically revise NESHAPs for source categories the agency has identified.[20] NESHAPs may include standards for major sources and for area sources, which are any sources not major.[21] Major source NESHAPs are based on the maximum achievable control technology (MACT), while EPA may use a different standard of generally available control technology for area sources.[22]

Major sources for NESHAPs are those that emit or have the potential to emit considering controls, in the aggregate, 10 tons per year or more of a single hazardous air pollutant or 25 tons per year or more of any

---

[17]The HAP provision also provided for EPA to submit a report to Congress in 1992 assessing the hazards to public health and the environment resulting from the emission of hydrogen sulfide associated with the extraction of oil and natural gas resources, and any recommendations, and directed EPA to, as appropriate, develop and implement a control strategy for such emissions using existing authorities. CAA § 112(n)(5), 42 U.S.C. § 7412(n)(5) (2012).

[18]Letter from Neil J. Carman, Sierra Club, et al, to Lisa Jackson, Administrator, EPA (Mar. 30, 2009). The petition states, among other things, that "[h]ealth studies support the need for EPA to list H₂S [hydrogen sulfide] under CAA section 112(b), especially since H2S's routine exposure effects – on a daily basis – are not addressed whatsoever under the accidental release provisions in section 112(r) of CAA." Id. at 1.

[19]Id. at 11.

[20]CAA § 112(d)(1), (d)(6), 42 U.S.C. §§ 7412(d)(1), (d)(6) (2012).

[21]Id. at § 7412(a)(1)-(2).

[22]Id. at § 7412(d)(5).

combination of HAPs.[23] Normally, the determination of a facility's potential to emit HAPs is based on the total of all activities at a facility, known as aggregation.[24] Under a unique provision of CAA, however, "emissions from any oil or gas exploration or production well (with its associated equipment) and emissions from any pipeline compressor or pump station shall not be aggregated with emissions from other similar units," to determine whether such units or stations are major sources of air pollution, or for other purposes under section 112 (e.g., the HAPs section).[25] Finally, facilities that do not contain a regulated unit (e.g., glycol dehydrator or covered storage vessel) are not subject to any requirement in the rule, even if they emit HAPs.[26]

Regarding the aggregation provisions, EPA officials explained that the agency has historically interpreted the statutory language to prohibit aggregation of HAP emissions from wells and associated equipment, meaning that each well and piece of associated equipment must be evaluated separately for purposes of determining major source status.[27] EPA has defined "associated equipment" in the regulations.[28] Officials said that EPA has not evaluated the significance of the aggregation prohibition and EPA's interpretation of it, such as its effect on the numbers of facilities that are or are not regulated as major sources and hence subject to MACT controls. Officials also said that it is likely that the effect of the aggregation provisions on well sites is smaller than its impact on downstream oil and gas production facilities where equipment tends to be larger and would be more likely to trigger MACT requirements if aggregated.

---

[23]Id. at § 7412(a)(1), 76 FR 52,741 (Aug. 23, 2011).

[24]40 C.F.R. § 63.2 (2012) (defining major source); see also 64 Fed. Reg. 32,610, 32,613 (June 17, 1999).

[25]CAA § 112(n)(4)(a), 42 U.S.C. § 7412(n)(4)(a) (2012).

[26]40 C.F.R. § 63.760(d) (2012).

[27]64 Fed. Reg. at 32,619.

[28]Id.; 40 C.F.R. § 63.761 (defining "associated equipment").

NESHAPs for Oil and Natural Gas Production Facilities

EPA originally promulgated the NESHAPs for Oil and Natural Gas Production source subcategory in two parts: the standard for major sources was issued in 1999, and the NESHAP for area sources in 2007.[29] In April 2012, EPA promulgated amendments to the NESHAPs for major sources.[30]

Major Sources

The NESHAPs for major sources apply to emission points of HAPs located at oil and natural gas production facilities (including wells, gathering stations, and processing plants) that are major sources.[31] Under this rule, in determining whether a well site's potential to emit HAPs equals or exceeds 10 tons per year (the major source threshold), only emissions from equipment other than wells or "associated equipment," may be aggregated; associated equipment is a defined term, and excludes glycol dehydrators and storage vessels.[32] In other words, emissions from wells are not aggregated; only emissions from glycol dehydrators and storage vessels at a site may be aggregated. Further, the rule exempts facilities exclusively handling and processing "black oil" and small oil and gas production facilities, including well sites, prior to the point of custody transfer.[33] EPA documents do not indicate the extent to which these exemptions have the effect of cancelling MACT requirements that would otherwise apply to oil and gas wells from unconventional deposits.

---

[29]64 Fed. Reg. 32,610 (June 17, 1999) (the notice also announced final rules for the category of natural gas transmission and storage facilities), 72 Fed. Reg. 26 (Jan. 3, 2007); 40 C.F.R. pt. 63 Subpt. HH.

[30]77 Fed. Reg. 49,490 (Aug. 16, 2012). The notice of the final rule was signed by the Administrator and a prepublication copy released to the public in April 2012.

[31]40 C.F.R. §§ 63.760(a)(1), 63.761 (2012).

[32]77 Fed. Reg. 49,490 , 49,501, 49,569 (Aug. 16, 2012) (revising 40 C.F.R. § 63.760, 63.761 to define major source such that "[f]or facilities that are production field facilities, only HAP emissions from glycol dehydration units and storage vessels shall be aggregated for a major source determination."). Storage vessels are defined as a tank or other vessel that is designed to contain an accumulation of crude oil, condensate, intermediate hydrocarbon liquids, or produced water and that is constructed primarily of nonearthen materials (e.g., wood, concrete, steel, plastic) that provide structural support. Id. at 49,569.

[33]64 Fed. Reg. 32,610, 32613 (June 17, 1999), 40 C.F.R. 63.760 (2012) (small sources are those, prior to the point of custody transfer, with a facility-wide actual annual average natural gas throughput less than 18.4 thousand standard cubic meters per day and a facility-wide actual annual average hydrocarbon liquid throughput less than 39,700 liters per day).

EPA headquarters officials did not know if any oil or gas wells were
NESHAP major sources prior to the April 2012 amendments, and EPA
officials in each of the four Regions we contacted were unaware of any
examples of oil and natural gas wells being regulated as major sources.
EPA officials noted that glycol dehydrators are more likely where there
are high pressure gas wells, such as in the Jonah-Pinedale area of
Wyoming. EPA officials said that a multiple pad well site in this area
would very likely be major for HAPs, except that any federally enforceable
standards are first applied to determine the potential emissions, and
Wyoming's presumptive best available control technology standards
would likely limit the emissions such that the potential to emit would be
reduced to area source levels. Analyses developed for the recent
amendments also do not identify if any well sites triggered the major
source NESHAPs prior to the amendments, but available data suggest
few well sites do so.[34]

The 1999 NESHAP for major sources has included the following emission
points that may be present at oil and gas wells: process vents on large
glycol dehydration units and storage vessels with potential for flash
emissions.[35] The standard requires reduction in HAPs emissions from
large glycol dehydration units and storage vessels with potential for flash
emissions through the application of air emission control equipment or
pollution prevention measures, or a combination of both.[36] The standards
require that all process vents on new and existing large glycol
dehydration units that are located at major HAPs sources be controlled.[37]
Glycol dehydration units subject to control must reduce emissions by 95
percent or more of HAPs, or to a benzene emission level less than 0.9
megagram (1 ton) per year.[38] For existing and new storage vessels with
the potential for flash emissions, the standard is to be equipped with a

---

[34]In addition to interviewing EPA officials in the Clean Air program and in four Regions,
GAO conducted sample searches on the EPA Air Facility System database.

[35]Large glycol dehydrators are those with an actual annual average natural gas flowrate
equal to or greater than 85 thousand standard cubic meters per day and actual annual
average benzene emissions equal to or greater than 0.90 Mg/yr. 77 Fed. Reg. 49,490 ,
49,568-69 (Aug. 16, 2012) (revising 40 C.F.R. §§ 63.761, 63.760(b)).

[36]64 Fed. Reg. 32,610, 32,613-14 (June 17, 1999).

[37]40 C.F.R. § 63.765 (2012).

[38]40 C.F.R. § 63.765 (2012), 64 Fed. Reg. 32,610, 32,613-14 (June 17, 1999).

cover vented through a closed vent system to a control device that recovers or destroys HAPs emissions with an efficiency of 95 percent or greater, or for combustion devices, reduces HAPs emissions to a specified outlet concentration.[39] However, these standards only apply at sites that are deemed major sources, and as outlined above, it appears likely that few well sites reach the key threshold emissions level.

The April 2012 amendments added one more emission source to the NESHAP major source rule: small glycol dehydrators.[40] These sources must meet a unit-specific limit for emissions of BTEX that is calculated using a formula in the rule based on the unit's natural gas throughput and gas composition.[41] Existing dehydrators have 3 years to comply, while new dehydrators must comply upon start-up.[42]

EPA had proposed another source to add to the NESHAPs: storage vessels without the potential for flash emissions, which are not regulated under the current rule. EPA did not include these sources in its final rule, stating that "the agency determined that it needs additional data in order to establish emission standards for this type of storage vessel."[43] In the proposed rule, EPA relied upon its original analysis of storage vessels that was used to determine the MACT standard in 1999, based on 1997 data.[44] Commenters criticized EPA's use of the 1999 analysis as outdated and not reflecting current technology in use for the vessels without potential for flash emissions and asserted that reliance on the old analysis failed to meet EPA's statutory obligations.[45] EPA stated "[i]n response to such comments, we have re-evaluated the proposed MACT standards and concluded that we need (and intend to gather) additional data on

---

[39]40 C.F.R. § 63.766 (2012).

[40]77 Fed. Reg. 49,490 ,49,568-71 (Aug. 16, 2012).

[41]Id. at 49,570-71.

[42]Id. at 49,568-60 (amending 40 C.F.R. 63.760(f)).

[43]Id. at 49,503; see also EPA, Summary of Requirements for Processes and Equipment at Natural Gas Well Sites.

[44]76 Fed. Reg. 52,738, 52,769 (Aug. 23, 2011).

[45]77 Fed. Reg. 49,490 , 49,503, 49,528-29 (Aug. 16, 2012).

these sources in order to analyze and establish MACT emission standards for this subcategory of storage vessels."[46]

In addition, the April 2012 amendments changed a key definition in the NESHAPs for determining major source status.[47] The effect of this change (i.e., revision to the definition of "associated equipment") is that emissions from *all* storage vessels and *all* glycol dehydrators now will be counted toward determining whether a facility is a major source under the NESHAP for Oil and Natural Gas Production. EPA documents do not indicate the extent to which the change in definition will result in additional oil and gas wells being subject to the MACT requirement.

Area Sources

CAA prohibits EPA from listing oil and gas production wells (with its associated equipment) as a specific "area source" category, unless the area source category is for oil and gas production wells located in any metropolitan statistical area or consolidated metropolitan statistical area with a population in excess of 1 million, and the EPA Administrator determines that emissions of HAPs from such wells present more than a negligible risk of adverse effects to public health.[48]

In 2007, EPA issued a NESHAP for oil and gas production facilities—which may include wells—that are area sources.[49] The area source rule regulations address emissions from one type of emission source at oil and gas production facilities: triethylene glycol dehydration units above specified throughput and benzene emission thresholds, which are the same thresholds as those used to define large dehydration units in the major source rule.[50] For area sources within defined control areas (areas of higher population density),[51] add-on controls or equivalent pollution prevention measures are required to achieve reduction of HAPs emissions by 95 percent, or alternately, to below the specified emission

---

[46]Id. at 49,503.

[47]Id. at 49,501, 49,569 (revising 40 C.F.R. § 63.761).

[48]CAA § 112(n)(4)(B), 42 U.S.C. § 7412(n)(4)(B) (2012).

[49]72 Fed. Reg. 26 (Jan. 3, 2007).

[50]40 C.F.R §§ 63.764(d), 63.765(a) (2012).

[51]EPA defines these control areas with reference to parameters used by the U.S. Census Bureau to identify densely settled areas. See 72 Fed. Reg. 26, 28 (2007).

limit for benzene.[52] For area sources outside of these control areas, an operational standard is required instead of an add-on control.[53]

Area sources are required to notify EPA that they are subject to the rule; additional information, including periodic reports, are required for area sources within a control area.[54] The area source notifications are sent to a specific EPA e-mail box. EPA does not track whether the facilities providing notification are well sites or other components of the oil and natural gas production sector, so it is difficult to determine to what extent oil and gas well sites are subject to the area source NESHAP.[55]

Regarding EPA's authority to establish an area source category for oil and gas wells in metropolitan statistical areas, if certain conditions are met, officials said that EPA has not considered doing so. They said that they have not analyzed well emissions in relation to location in or outside a metropolitan statistical area, and that if the agency were to consider developing an area source within metropolitan statistical areas, they would need to conduct a new data collection effort.

Other NESHAPs

In addition, EPA has promulgated other NESHAPs, the applicability of which to oil and gas well sites depends upon the particular equipment—and factors such as capacity or emission rate—used at a well site. Although some published materials suggest several NESHAPs may apply, based on discussions with EPA, the primary NESHAP that officials believe could apply at oil and gas well sites is the Boilers and Process Heaters NESHAP for major sources.[56]

---

[52]72 Fed. Reg. 26, 28 (Jan. 3, 2007), 40 C.F.R §§ 63.764(d), 63.765(a) (2012).

[53]40 C.F.R § 63.764(d) (2012).

[54]40 C.F.R. § 63.775(c) (2012); see also 72 Fed. Reg. 26, 30 (Jan. 3, 2007).

[55]For example, GAO searched EPA's Air Facility System database to identify 40 C.F.R. pt. 63 Subpt. HH MACT area sources with the Standard Industrial Classification code for oil and gas extraction. Some of the facilities identified in the search results have "well" in the facility name, but may include other facilities downstream of the well pad; the database does not have information to distinguish the well sites among the facilities.

[56]40 C.F.R. pt. 63 Subpt. DDDDD, §§ 63.7480-7575 (2012).

The major source rule for boilers and process heaters has an unusual feature in that, to determine applicability of the rule, it references whether or not an oil and gas production facility falls within the major source definition under the NESHAPs for Oil and Gas Production Facilities (subpart HH).[57] If an oil and gas well were a major source under the Oil and Gas NESHAP, then any boilers or process heaters with heat input of 10 million British thermal units (BTU) per hour are subject to emission limitations requirements, and any smaller heaters are subject to work standards, under the Boiler NESHAP. These requirements differ from those in the NESHAP for Oil and Gas Production Facilities by, among other things, imposing limits for other pollutants, such as particulate matter, hydrogen chloride, mercury, carbon monoxide, and dioxins/furans, depending on the type of unit.[58] Officials stated that some glycol dehydrators at well sites could be over the trigger heat input and would be subject to the Boiler NESHAP requirements if the oil and gas site were a major source subject to the rule. As noted above, it is not known how many, if any, well sites are major sources.

Where a gas well has a compressor, the compressor engine may be subject to standards for stationary engines.[59] EPA did not have available information on the extent to which these engines are present at well sites and, if so, whether they fall under these rules, which are based on equipment and are not specific to the oil and gas industry.

New Source Performance Standards

EPA promulgates NSPS, which are generally applicable to (1) new or reconstructed facilities and (2) facilities that have undergone modification—that is, any physical change in, or change in the method of operation of, a facility which increases the amount of any air pollutant emitted by such source or which results in the emission of any air pollutant not previously emitted.[60] These rules are implemented by EPA

---

[57] 40 C.F.R. § 63.7485, 76 Fed. Reg. 15,608, 15,619 (Mar. 21, 2011).

[58] 40 C.F.R. § 63.7500, pt. 63 Subpt. DDDDD Tables 1-2 (Boiler NESHAP); cf. 40 C.F.R. pt. 63 Subpt. HH App. Table 1, 64 Fed. Reg. 32,610 (June 17, 1999) (Oil and Gas NESHAP for major sources). The NESHAP for Oil and Gas Production Facilities is focused on BTEX and n-hexane.

[59] 40 C.F.R. pt. 63, Subpt. ZZZZ (Stationary Reciprocating Internal Combustion Engines).

[60] CAA § 111(a)(2), (4), 42 U.S.C. § 7411(a)(2), (4). (2012).

GAO-12-874 Unconventional Oil and Gas Development

or by states through delegation.[61] For the oil and gas production industry, the NSPS primarily regulates VOCs (as an ozone precursor).

In 1985, EPA promulgated NSPS for the oil and gas industry focused on natural gas processing plants, but did not include any standards for emissions from preprocessing production activities.[62] EPA has recently promulgated such standards for some production emissions, notably completion and recompletion of certain hydraulically fractured gas wells.[63] In addition, some other generally applicable standards for certain equipment may apply at oil and gas well sites.

April 2012 Amendments to NSPS

In April 2012, EPA promulgated amendments to the NSPS for the Oil and Gas sector, including new standards applicable to the production source category.[64] The new standards were issued pursuant to a 2010 consent decree that settled a challenge brought by environmental groups over EPA's failure to conduct required reviews of the existing standards.[65] Following publication of the new rules in August 2012, an industry group petitioned EPA to reconsider certain aspects of the new rules.[66]

The new standards include, for the first time, standards to reduce emissions from certain activities at natural gas wells.[67] In particular, the new standards establish operational standards applicable to selected

---

[61]Id. at § 7411(c).

[62]40 C.F.R. pt. 60, subparts KKK, LLL.

[63]77 Fed. Reg. 49,490 , 49,542-67 (Aug. 16, 2012) (adding 40 C.F.R. pt. 60 subpt. OOOO, consisting of §§ 60.5360 to 60.5430).

[64]Id.

[65]See Consent Decree, Document 25, and Third Stipulation of the Parties to Modify Consent Decree, Document 28, Wildearth Guardians et al. v. Jackson, No. 1:09-cv-00089-CKK (Dist. D.C. 2011). See also CAA §§ 111(b)(1)(B), 112(d)(6), (f)(2), 42 U.S.C. §§ 7411(b)(1)(B), 7412(d)(6), (f)(2) (2012).

[66]American Petroleum Institute, Request for Administrative Reconsideration and an Administrative Stay of Targeted Elements of EPA's Final Rule "Oil and Natural Gas Sector: New Source Performance Standards and National Emission Standards for Hazardous Air Pollutants Reviews," Aug. 16, 2012.

[67]77 Fed. Reg. 49,490, 49,497-98, 49,543 (Aug. 16, 2012) (adding, e.g., 40 C.F.R. § 60.5365).

completions and recompletions of natural gas wells, with variable
implementation dates as described in table 10. These practices are
designed to capture emissions from flowback from hydraulically fractured
wells, and reduce VOC emissions. EPA's regulatory impact analysis
estimated that the rules will apply to about 9,700 new wells per year, and
to about 1,200 existing wells being recompleted per year.[68] Of these,
EPA's analysis estimates that nearly 9,400 wells will be required to use
"green completion" techniques to capture and treat flowback emissions so
that the captured natural gas can be sold or otherwise used, while the
remainder will use completion combustion.

Table 10: NSPS for Natural Gas Wells, by Well Subcategory

Natural gas well subcategory	Completion date	Green completions (routing gas to the flow line)	Combustion of flowback emissions	Duty to minimize releases to the air (flowback and recovery)	Well logs, compliance demonstration, records and reporting
Provisions[b]		40 C.F.R. § 60.5375(a)(1)-(2)	40 C.F.R. § 60.5375(a)(3)	40 C.F.R. § 60.5375(a)(4)	40 C.F.R. § 60.5375(b)-(e)
New wells constructed after Aug. 23, 2011	Completions with hydraulic fracturing before Jan. 1, 2015		√	√	√
	Completions with hydraulic fracturing after Jan. 1, 2015	√	√ (only for emissions that cannot be directed to the flow line)	√	√
Wells existing as of Aug. 23, 2011	Completions with hydraulic fracturing before Jan. 1, 2015		√	√	√
	Completions with hydraulic fracturing after Jan. 1, 2015	√	√ (only for emissions that cannot be directed to the flow line)	√	√
	Alternative to avoid new source status before Jan. 1, 2015[a]	√	√ (only for emissions that cannot be directed to the flow line)	√	√

[68]See EPA, *Regulatory Impact Analysis: Final New Source Performance Standards and Amendments to the National Emissions Standards for Hazardous Air Pollutants for the Oil and Natural Gas Industry* at 3-12 (April 2012). EPA projected these numbers of affected wells for the year 2015, when the rule will be fully in effect, and the number of wells affected may actually vary from year to year. At the time of the proposed rule, EPA estimated that over 20,000 completions and recompletions annually would be subject to the proposed requirements. 76 Fed. Reg. 52,738, 52,747 (Aug. 23, 2011).

Natural gas well subcategory	Completion date	Green completions (routing gas to the flow line)	Combustion of flowback emissions	Duty to minimize releases to the air (flowback and recovery)	Well logs, compliance demonstration, records and reporting
Wildcat and delineation wells		√		√	√
Other low pressure wells		√		√	√

Source: GAO analysis of EPA documents.

[a]Existing wells (constructed before Aug. 23, 2011) that are completed after fracturing can avoid being within the definition of a "modified" source by voluntarily implementing the green completion provisions. According to EPA, this would allow the owners/operators to avoid state permit requirements in some states. See EPA, Summary of Requirements for Processes and Equipment at Natural Gas Well Sites (April 2012).

[b]Provisions listed are to code sections; see 77 Fed. Reg. at 49,543-54.

Additionally, to reduce VOC emissions, the April 2012 rule establishes standards including those for, as relevant to gas well sites, gas-driven pneumatic controller devices and storage vessels, subject to thresholds.[69] According to EPA documents, over 13,600 pneumatic controllers will be affected, but it is not clear the extent to which these are located at well sites.[70] Similarly, EPA documents estimate that 304 storage vessels annually will trip the threshold of 6 tons per year of VOC and thus be subject to the rule, and EPA officials expect most of these storage vessels will be located at wells.[71]

When asked about the potential increased burden of the amended NSPS rules, officials said that it was not clear whether the rule would result in more or fewer CAA-related permits. For example, the applicability of

[69]For pneumatic controllers, an affected source is a single continuous bleed natural gas-driven pneumatic controller operating at a natural gas bleed rate greater than 6 standard cubic feet per hour located between the wellhead and the point of custody transfer to the natural gas transmission and storage segment and not located at a natural gas processing plant or oil pipeline. 77 Fed. Reg. at 49,543 (adding 40 C.F.R. § 60.5365(d)(i)).For storage vessels, the threshold for coverage is 6 tons per year of VOC. 77 Fed. Reg. at 49,543, 49,556 (adding 40 C.F.R. §§ 60.5365(e), 60.5430).

[70]EPA, Regulatory Impact Analysis: Final New Source Performance Standards and Amendments to the National Emissions Standards for Hazardous Air Pollutants for the Oil and Natural Gas Industry (April 2012) at 3-12.

[71]EPA also regulated centrifugal and reciprocating compressors at downstream sites, but those located at a well site, or an adjacent well site and servicing more than one well site, are not covered. 77 Fed. Reg. at 49,543 (adding 40 C.F.R. § 60.5365(b), (c)).

NSPS may trigger a state requirement to get a construction permit or other type of permit. These permits may be triggered by, among other things, a facility's "potential to emit" that is calculated assuming all federally enforceable controls are in place. Officials said that the NSPS, which are federally enforceable requirements, will reduce actual emissions and thus could reduce the number of facilities that trigger the requirement for these state permits. In the new rule, EPA generally exempted covered facilities from the obligation to obtain a Title V operating permit.[72]

Other NSPS

EPA has issued equipment-focused NSPS for certain equipment that may be used at oil and gas well sites. These include NSPS for Volatile Organic Liquid Storage Vessels (Including Petroleum Liquid Storage Vessels).[73] These standards apply to such tanks with a capacity greater than or equal to 75 cubic meters that is used to store volatile organic liquids and that were built, reconstructed, or modified after July 23, 1984. Tanks attached to trucks and other mobile vehicles are excluded.[74] EPA officials said that, while there are tanks at well sites, they are often smaller than the threshold in this rule. Specifically, while the standards apply to tanks greater than 75 cubic meters (about 475 barrels, according to EPA), an individual tank typically found at oil and gas sites is often between 250 – 400 barrels, hence avoiding coverage under this rule.

Other NSPS that have been identified as potentially relevant include those for gas turbines and steam generators.[75] EPA officials said, however, that typical activity at well sites is not enough to trigger thresholds for coverage under this rule, either.

---

[72]77 Fed. Reg. at 49,543 (adding 40 C.F.R. § 60.5370(c)) (provided the facility is not otherwise required by law to obtain a Title V operating permit or new source review permit).

[73]40 C.F.R. pt. 60, subpt. Kb (2012).

[74]40 C.F.R. § 60.110b(d)(3) (2012).

[75]40 C.F.R. pt. 60, subpt. D to Dc (2012). See also EPA, Sector Notebook Project: Oil and Gas Extraction, at 100 (2000).

New Source Review

CAA New Source Review (NSR) provisions require a source to obtain a permit and undertake other obligations to control its emissions of air pollution prior to construction of a new source or modification of an existing stationary source.[76] However, NSR only applies if the construction project results in actual emissions or the potential to emit regulated air contaminants[77] at or above certain threshold levels established in the NSR regulations. For a new source, NSR is triggered only if the emissions would cause the source to qualify as major. For an existing major source making a modification, NSR is triggered only if the modification will result in a significant increase in emissions and a significant net emissions increase of that pollutant. Relevant to NSR, the emission profile for oil and gas wells would include hydrogen sulfide and VOCs, among others. In most areas, states implement the NSR permitting programs.

The major NSR program is actually composed of the following two separate programs:

- Nonattainment NSR applies to emission of specific pollutants from sources located in areas designated as nonattainment for those pollutants because they do not meet the pollutant-specific national ambient air quality standards.

- Prevention of Significant Deterioration (PSD) applies to emissions of all other regulated pollutants from sources located in attainment areas where such standards are met or in areas unclassifiable for such standards.

For PSD, the major source threshold is generally 250 tons per year of any regulated air pollutant.[78] Determining whether a facility is a major source, together with identifying which emissions should be included in doing so is guided by the process as for Title V permits, discussed below.[79] While a

---

[76]CAA §§ 165(a), 173(a), 42 U.S.C. §§ 7475(a), 7503(a) (2012); see generally CAA §§ 160-169, 171-189, 42 U.S.C. §§ 7470-7479 (Prevention of Significant Deterioration), 7501-7513 (NSR nonattainment) (2012).

[77]Potential to emit generally means the maximum capacity of a stationary source to emit any air pollutant under its physical and operational design, taking into account any federally enforceable limitations.

[78]CAA §§ 169(1), 165(a), 42 U.S.C. §§ 7479(1), 7475(a) (2012).

[79]40 C.F.R. § 52.21(b)(5)-(6) (2012).

1993 EPA report appeared to suggest that most oil and gas extraction wells would not likely be subject to PSD regulations based on the applicability criteria,[80] the specific determination of which emission units, including wells, must be included in determining whether a source is major (source aggregation) involves a case-by-case, fact-specific analysis.[81] For nonattainment NSR, the major source threshold ranges from 100 tons per year down to 10 tons per year depending on the severity of the air quality problem where the source is located and the specific pollutant at issue.[82] To be a major source under nonattainment NSR, the source must emit or have a potential to emit above the major source level set for the specific regulated air pollutant (or its precursor) for which the area is designated nonattainment.[83] With respect to nonattainment NSR, EPA officials stated that some large wells in nonattainment zones could be major sources standing alone because of low emission thresholds in certain areas; as noted above, such thresholds could be as low as 10 tons per year in the most severe nonattainment areas, versus 250 tons per year in attainment areas.

Title V Operating Permits

Relevant to oil and gas production, CAA generally requires Title V permits for the operation of

- any major source[84] determined based on the facility's actual emissions or "potential to emit;"

- any source, including a nonmajor source, subject to a NSPS;

- any source, including an area source, subject to a NESHAP, among others; and

- any source required to obtain a PSD or NSR permit.[85]

---

[80]EPA, Report to Congress on Hydrogen Sulfide Emissions Associated with the Extraction of Oil and Gas, EPA-453-R-93/045 at IV-37 (1993).

[81]Gina McCarthy, AA Office of Air and Radiation, Memorandum, "Withdrawal of Source Determinations for Oil and Gas Industries" (Sept. 22, 2009).

[82]40 C.F.R. § 51.165 (2012).

[83]40 C.F.R. § 51.165 (2012).

[84]Major source for Title V purposes is defined more broadly than for NESHAP purposes. Cf. CAA § 501(2) to § 112(a)(1), 42 U.S.C. § 7661(2) to § 7412(a)(1) (2012).

Thus, whether a Title V permit is required depends on whether the source (1) is subject to one of these other requirements, unless EPA has exempted the particular area sources or nonmajor sources from the Title V permit requirement,[86] or (2) meets the emissions thresholds for a major source. Title V permits for a major source must include all applicable requirements for all relevant emission units in the major source.[87] Title V permits for nonmajor sources must include all applicable requirements applicable to emissions units that caused the source to be subject to the Title V permitting requirements.[88] Title V permits may need to add monitoring, reporting, or other requirements but generally do not add new emissions control requirements (rather they consolidate requirements from throughout CAA programs and contain conditions to assure compliance with such requirements). According to EPA officials, the permits help operators and the public to understand what the requirements are for compliance with CAA and help assure compliance with such requirements.

Title V permits are generally issued by states and, in some instances, EPA Regional offices. As of August 2012, EPA officials were unaware of any Title V permits issued solely on the basis of oil and gas well site emissions alone. EPA officials stated that some oil and gas well sites have adopted federally enforceable emissions limits such that the sites do not need a Title V permit, which they would otherwise have triggered. In addition, EPA identified a March 2012 case in which a state environmental agency alleged, among other things, that an oil and gas production site had VOC emissions of over 600 tpy, which would require a Title V permit. The operator disputed the violations but agreed to submit an application for a Title V permit.

---

[85]CAA § 501(2), 42 U.S.C. § 7661(2) (2012), 40 C.F.R. §§ 70.3(a), 70.5(a)(1)(ii), 71.3(a), 71.5(a)(1)(ii) (2012).

[86]EPA has the authority to exempt through rulemaking area sources or non-major sources from the Title V permit requirement if they are not otherwise subject to Title V. CAA § 502(a), 42 U.S.C. §7661a(a) (2012). Major sources under Title V may not be exempted.

[87]40 C.F.R. §§ 70.3(c)(1), 71.3(c)(1) (2012).

[88]40 C.F.R. §§ 70.3(c)(2) 71.3(c)(2) (2012).

Source Determinations and Aggregation Issues for Title V and NSR

Applicable to both NSR and related determinations for Title V, EPA regulations specify three factors that must be met in source determinations—whether the emissions points are under common control, belong to the same major industrial grouping,[89] and are located on contiguous or adjacent properties.[90] Thus, in contrast to the NESHAPs, in determining whether significance thresholds for emissions are met for purposes of NSR or Title V, EPA and states must aggregate VOC emissions from oil and gas well sites that are both (1) contiguous or adjacent and (2) under common control. To determine whether a source meets the emissions thresholds for a Title V or NSR major source designation, EPA applies these regulatory criteria to evaluate whether to aggregate oil and gas production wells with other emission sources.[91] Specifically, permitting authorities (EPA or authorized states or local authorities) have in particular matters, on a case-by-case basis, aggregated emissions from facilities to determine major sources, for purposes of Title V operating permits or NSR. Determining when emissions must be aggregated is a fact-based inquiry that is made by permitting authorities on a case-by-case basis. While authorized states are typically responsible for making source determinations, EPA headquarters has stated that Regional offices should continue to review and comment on source determinations to assure consistency with regulations and historical practice.[92] In addition, EPA Regions may be responsible for source determinations in areas where they are responsible for permitting.

Aggregation of emissions from the oil and gas industry generally, including production facilities, has received recent attention.[93] For

---

[89]All oil and gas emission units are in the same major industrial grouping.

[90]40 C.F.R. §§ 71.2, 70.2 (2012) (definition of major source for Title V permitting), §§ 51.165(a)(1)(i)-(ii), (iv) (definition of major stationary source for NSR), 51.166(b)(5)-(6), 52.21(b)(5)-(6) (definitions of stationary source for NSR) (2012).

[91]See, e.g., Letter, Cheryl L. Newton, EPA Air and Radiation Division, to Scott Huber, Summit Petroleum Corporation (Oct. 18, 2010) (aggregating gas wells, a sweetening plant, and associated flares as constituting a single source for purpose of Title V permitting).

[92]Gina McCarthy, AA Office of Air and Radiation, Memorandum, "Withdrawal of Source Determinations for Oil and Gas Industries" at 2 (Sept. 22, 2009).

[93]See, e.g., Roger Martella et al, Aggregation of Oil and Gas Wells Under the Clean Air Act: New Horizons from EPA and the Courts, Natural Resources and Environment (Fall 2011); http://www.natlawreview.com/article/oil-gas-activities-agencies-are-air-air-regulations

example, in 2007, EPA provided guidance on how to evaluate aggregation in source determinations for the oil and gas industry.[94] EPA later withdrew this industry-specific guidance and emphasized that source determinations in this industry were governed by the existing regulations, the existing interpretations of them, and need for case-specific application of the regulations in each permitting action.[95]

Aggregation of oil and gas facilities—including wells—has also been the subject of litigation. For example, a gas production company challenged EPA's determination that its gas wells, sweetening plant, and associated flares constitute a single source for purposes of Title V permitting, in the United States Court of Appeals for the Sixth Circuit.[96] In a recently issued decision, the court vacated and remanded EPA's determination, finding that EPA improperly considered the functional interrelatedness of the sweetening plant and the wells in determining that those points were adjacent under the regulations. Another example of a challenge relates to a citizen petition filed in EPA Region 8.[97] The state of Colorado issued a Title V permit renewal to Anadarko's Frederick Compressor Station for a natural gas processing station but did not aggregate the station with natural gas wells for purposes of Title V and also did not include a requirement of PSD because the emission threshold was not met without aggregation. A citizen group subsequently filed a Title V petition with EPA, seeking an objection to the permit because natural gas wells were not aggregated with the processing station. After EPA issued a Title V order finding that the petition had not demonstrated that aggregation was required, the citizen group challenged the EPA decision in the United States Court of Appeals for the Tenth Circuit.[98] In another matter, EPA Region 8 issued a Title V permit to BP America Production Company's Florida River Compression Station Facility and decided not to aggregate

[94]See Id.; William Wehrum, Memorandum, "Source Determinations for Oil and Gas Industries" (Jan. 12, 2007).

[95]Gina McCarthy, AA Office of Air and Radiation, Memorandum, "Withdrawal of Source Determinations for Oil and Gas Industries" (Sept. 22, 2009).

[96]See *Summit Petroleum Corp. v. EPA*, Nos. 09-4348, 10-4572, slip op. (6th Cir. August 7, 2012).

[97]See EPA, Order Responding To Petitioners' Request That The Administrator Object To Issuance Of A State Operating Permit, Petition Number: VIII-2010-4 (Feb. 2, 2011).

[98]*WildEarth Guardians et al. v. Jackson*, No.11-9527 (10th Cir).

oil and gas activities with the compressor station in determining the source. A citizen group appealed this decision to the Environmental Appeals Board.[99] Both citizen group challenges were ultimately dismissed after the parties engaged in a dispute resolution process. EPA entered settlements with the citizen group and agreed to undertake a pilot program for the purpose of studying, improving, and streamlining source determinations in the oil and gas industry in new or renewal Title V permits for which EPA Region 8 is the initial Title V permitting authority.[100]

In sum, several recent disputes over aggregation of oil and gas facilities involve whether or not well emissions should be aggregated; however, whether or not well emissions are aggregated for Title V or PSD purposes generally would not affect other federal requirements for emission controls at well sites.

Greenhouse Gas Reporting Rule

In 2009, EPA promulgated the Greenhouse Gas Reporting Rule, providing a framework for the greenhouse gas reporting program and establishing requirements for some source categories.[101] According to EPA, the goals of the program are to obtain data that are of sufficient quality that they can be used to support a range of future climate change policies and regulations; to balance the rule coverage to maximize the amount of emissions reported while minimizing reporting from small emitters; and to create reporting requirements that are consistent with existing programs by using existing estimation and reporting methodologies to reduce reporting burden, where feasible.[102]

EPA subsequently issued and amended a rule to implement the program for the category of Petroleum and Natural Gas Systems, including oil and

---

[99]*In re BP America Production Co., Florida River Compression Facility*, Environmental Appeals Board Appeal No. CAA 10–04.

[100]*Wildearth Guardians v. EPA*, Settlement Agreement at 2-3, Docket No. 11-9527 (10th Circuit 2011). See also 76 Fed. Reg. 71,027 (Nov. 16, 2011).

[101]74 Fed. Reg. 56,260 (Oct. 30, 2009), 40 C.F.R. pt. 98.

[102]See, e.g., EPA, Economic Impact Analysis for the Mandatory Reporting of Greenhouse Gas Emissions Under Subpart W Supplemental Rule (GHG Reporting), Final Report 2-1 – 2-2 (2009).

gas wells.[103] According to EPA, oil and gas well sites may contain sources of greenhouse gas emissions including: (1) combustion sources, such as engines used on-site and which typically burn natural gas or diesel fuel, and (2) process sources, such as equipment leaks and vented emissions.[104] The process sources include pneumatic devices, dehydrators, and compressors. EPA has identified the onshore production subcategory as the largest segment for equipment leaks and vented and flared emissions in the petroleum and natural gas system source category.[105]

The rule requires petroleum and natural gas facilities—including oil and gas well sites—that emit 25,000 metric tons or more of carbon dioxide equivalent[106] per year to report certain data to EPA. Specifically, oil and gas production facilities are to report annual emissions of carbon dioxide, methane, and nitrous oxide from

- equipment leaks and venting,

- gas flaring, and

- stationary and portable combustion.[107]

Reporting is to begin in September 2012, for calendar year 2011.[108]

---

[103]75 Fed. Reg. 74,458 (Nov. 30, 2010), 76 Fed. Reg. 59,533 (Sept. 27, 2011), 76 Fed. Reg. 73,886 (Nov. 29, 2011), 76 Fed. Reg. 80,554 (Dec. 23, 2011); 40 C.F.R. pt. 98 Subpt. W. See also http://www.epa.gov/climatechange/emissions/subpart/w.html

[104]EPA, Quantifying Greenhouse Gas Emissions from Key Industrial Sectors in the United States 12-2 (Working Draft May 2008).

[105]EPA, Economic Impact Analysis for the Mandatory Reporting of Greenhouse Gas Emissions Under Subpart W Supplemental Rule (GHG Reporting), Final Report 1-10 (2009).

[106]Carbon dioxide equivalent measures the total greenhouse gases, accounting for the relative ability of each greenhouse gas to trap heat in the atmosphere, known as the global warming potential.

[107]40 C.F.R. § 98.232(c), (c)(12)-(13) (well flaring during testing and production), (c)(21) (leaks), (c)(22) (combustion) (2012).

[108]76 Fed. Reg. 73,886, 73,889, 73,899 (Nov. 29, 2011) (amending 40 C.F.R. § 98.3).

For purposes of this rule, onshore petroleum and natural gas production is defined to include all equipment on a single well pad or associated with a single well pad (including but not limited to compressors, generators, dehydrators, storage vessels, and portable non-self-propelled equipment which includes well drilling and completion equipment, workover equipment, gravity separation equipment, auxiliary non-transportation-related equipment, and leased, rented or contracted equipment or storage facilities), used in the production, extraction, recovery, lifting, stabilization, separation, or treating of petroleum and/or natural gas (including condensate).[109] Moreover, the rule defines an onshore oil and gas production facility as including all oil or gas equipment on or associated with a well pad and carbon dioxide enhanced oil recovery operations that are under common ownership or control and that are located in a single hydrocarbon basin; thus, for example, where multiple wells are owned or operated by the same person or entity in a single basin, the owner or operator is to report well data collectively for each hydrocarbon basin.[110] EPA estimated that this facility definition for onshore petroleum and natural gas production will result in 85 percent GHG emissions coverage of this industry segment,[111] and EPA documents estimate that emissions from approximately 467,000 onshore wells are covered under the rule.[112]

Accidental Releases

Section 112(r) of CAA establishes the chemical accidental release prevention program applicable to specifically listed "regulated substances," as well as other extremely hazardous substances. This provision, among other things, required EPA to publish regulations and guidance for chemical accident prevention at facilities using substances

[109]75 Fed. Reg. 74,458, 74,461 (Nov. 30, 2010), 40 C.F.R. § 98.230(a)(2) (2012).

[110]40 C.F.R. §§ 98.238, 98.236(c)(10)-(11) (2012) ("Facility with respect to onshore petroleum and natural gas production for purposes of reporting under this subpart and for the corresponding subpart A requirements means all petroleum or natural gas equipment on a single well-pad or associated with a single well-pad and CO2EOR operations that are under common ownership or common control including leased, rented, or contracted activities by an onshore petroleum and natural gas production owner or operator and that are located in a single hydrocarbon basin as defined in §98.238. Where a person or entity owns or operates more than one well in a basin, then all onshore petroleum and natural gas production equipment associated with all wells that the person or entity owns or operates in the basin would be considered one facility.").

[111]75 Fed. Reg. 74,458, 74,467 (Nov. 30, 2010).

[112]75 Fed. Reg. 74,458, 74,479 (Nov. 30, 2010).

that pose the greatest risk of harm from accidental releases;[113] the resulting regulatory program is known as the Risk Management Program. In conjunction with the program, EPA was required to promulgate a list of at least 100 substances which, in the case of an accidental release, are known to cause or may reasonably be anticipated to cause death, injury, or serious adverse effects to human health or the environment, and to periodically review the list.[114] Among others, hydrogen sulfide is included on the list of regulated substances.[115] Section 112(r) also established the Chemical Safety Board;[116] and the general duty for owners and operators of facilities to take steps to prevent accidental releases of the listed and other extremely hazardous substances, among other things.[117]

Accidental Release Prevention (Risk Management Program)

Whether and the extent to which a facility is subject to the Risk Management Program requirements depends on the regulated substances present and their quantities, the processes, and the presence of receptors. Generally, the regulation requires, for covered processes, a three-part program including (1) a hazard assessment; (2) a prevention program that includes safety procedures and maintenance, monitoring, and employee training measures; and (3) an emergency response program.[118]

EPA's list of regulated substances and their thresholds for the Risk Management Program was initially established in 1994 and has been revised several times.[119] As amended, the following chemicals that may be found at oil and gas sites are excluded from threshold determinations:

[113]CAA § 112(r)(7), (r)(2)(A), 42 U.S.C. § 7412(r)(7), (r)(2)(A) (2012) (defining accidental release as "an unanticipated emission of a regulated substance or other extremely hazardous substance into the ambient air from a stationary source.").

[114]Id. at § 7412(r)(3).

[115]40 C.F.R. § 68.130 (2012).

[116]CAA § 112(r)(6), 42 U.S.C. § 7412(r)(6), (6)(C) (2012).

[117]Id. at § 7412(r)(1).

[118]40 C.F.R. § 68.12 (2012).

[119]63 Fed. Reg. 640 (Jan. 6, 1998); 65 Fed. Reg. 13,243, 13,244 (Mar. 13, 2000).

- naturally occurring hydrocarbon mixtures, which include any combination of the following: condensate, crude oil, field gas, and produced water (defined as water extracted from the earth from an oil or natural gas production well, or that is separated from oil or natural gas after extraction),[120]

- regulated substances in gasoline, when in distribution or related storage for use as fuel for internal combustion engines,[121] and

- a flammable substance when the substance is used as a fuel.[122]

Regarding the exemption of naturally occurring hydrocarbon mixtures prior to entry into a processing plant or refinery, EPA explained at the time that the agency believed they do not warrant regulation, noting that the general duty clause would apply when site-specific factors make an unlisted chemical extremely hazardous.[123] In addition, EPA stated that, for naturally occurring hydrocarbons and for regulated substances in gasoline, a key consideration was EPA's original intent to exempt flammable mixtures that do not meet a preexisting standard—the National Fire Protection Association flammability hazard rating of 4. EPA has also explained that this rating reflects the potential to result in vapor cloud explosions and boiling liquid expanding vapor explosions, which it found pose the greatest potential hazard from flammable substances to the public and environment.[124]

Regarding flammable substances used as fuel, EPA had originally included chemicals on the flammable substances list based on the National Fire Protection Association flammability hazard rating, regardless of their use. After promulgating Risk Management Program regulations, EPA became aware that certain small commercial sources were subject to the requirements because they used propane or other fuels, so it initiated a rulemaking to create an exemption. In addition, a propane gas industry association had challenged the Risk Management

---

[120]40 C.F.R § 68.115(b)(2)(iii), 68.3 (2012); 63 Fed. Reg. 640, 641 (Jan. 6, 1998).

[121]40 C.F.R § 68.115(b)(2)(ii) (2012); 63 Fed. Reg. 640, 641 (Jan. 6, 1998).

[122]40 C.F.R § 68.126 (2012).

[123]63 Fed. Reg. 640, 641 (Jan. 6, 1998).

[124]65 Fed. Reg. 13,243, 13,245 (Mar. 13, 2000).

Program rule. In this context, the Chemical Safety Information, Site
Security and Fuels Regulatory Relief Act prohibited EPA from listing
flammable substances used as fuel, solely because of their explosive
potential.[125] EPA then revised the regulation, adding the exemption to
comply with the act.[126]

The regulated chemicals present at oil and gas well sites include
components of natural gas (such as butane, propane, methane, and
ethane), but these are exempt from the threshold determination of a
facility subject to the Risk Management Program when present in
"naturally occurring hydrocarbon mixtures."[127] If an oil or gas well site
nonetheless uses or stores some of the regulated chemicals not
encompassed by the exemptions, it could trigger the risk management
requirements.

General Duty to Prevent Accidental Releases

Section 112(r) also provides, in relevant part:

> The owners and operators of stationary sources producing, processing, handling or
> storing such substances have a general duty ...to identify hazards which may result
> from such releases using appropriate hazard assessment techniques, to design and
> maintain a safe facility taking such steps as are necessary to prevent releases, and to
> minimize the consequences of accidental releases which do occur.[128]

Known as the "general duty clause," the provision is analogous to a
negligence standard, according to EPA officials. In other words, if there is
a known risk and a way to mitigate it, then the operator should conduct
risk mitigation. As explained in an EPA report, "responsibilities include the
conduct of appropriate hazard assessments and the design, operations,

---

[125]Pub. L. No. 106–40 § 2,113 Stat. 207 (1999) (Section 2 of the Act immediately
removed EPA's authority to "list a flammable substance when used as a fuel or held for
sale as a fuel at a retail facility * * * solely because of the explosive or flammable
properties of the substance, unless a fire or explosion caused by the substance will result
in acute adverse health effects from human exposure to the substance, including the
unburned fuel or its combustion byproducts, other than those caused by the heat of the
fire or impact of the explosion."

[126]65 Fed. Reg. at 13,247.

[127]40 C.F.R. § 68.115(b)(2)(iii) (2012).

[128]CAA § 112(r)(1), 42 U.S.C. § 7412(r)(1) (2012).

and maintenance of a safe facility," as well as release mitigation and community protection.[129] EPA officials noted that industry standards (such as from the American National Standards Institute or the American Petroleum Institute) and fire codes are used in determining the duty of care. EPA has published Chemical Safety Alerts to advise the regulated community of its general duty clause obligations.[130]

The general duty clause applies to sources handling or storing substances listed by EPA in the Risk Management Program regulations or any other extremely hazardous substance, without a threshold. EPA headquarters officials said that, conceivably, the general duty clause would apply to every single well but stated that it would be in EPA Regions' discretion where and when to use the general duty clause to conduct inspections. In some Regions, EPA has conducted inspections of gas well sites to enforce the general duty clause, including identifying noncompliance with certain safety standards. EPA Regional officials said that they use infrared video cameras to conduct inspections to identify leaks of methane from storage tanks or other equipment at well sites. For example, EPA Region 6 officials said they have conducted 45 inspections at well sites since July 2010 and issued 10 administrative orders related to violations of CAA general duty clause. EPA officials said that all well sites are required to comply with the general duty clause but that EPA prioritizes and selects sites for inspections based on risk.

Imminent and Substantial Endangerment Authority Respecting Accidental Releases

Section 112(r) also provides EPA with the authority to issue orders as may be necessary to protect the public health when the EPA Administrator determines that there may be an imminent and substantial endangerment to human health or welfare or the environment because of an actual or threatened accidental release of a regulated substance.[131]

---

[129]EPA, *Report to Congress on Hydrogen Sulfide Emissions Associated with the Extraction of Oil and Gas*, EPA-453-R-93/045 at IV-35 (1993).

[130]See http://www.epa.gov/oem/publications.htm#alerts

[131]CAA § 112(r)(9)(A), 42 U.S.C. § 7412(r)(9)(A) (2012).

Chemical Safety Board	The Chemical Safety Board, established by section 112(r), is charged with investigating and publicly reporting on accidental releases resulting in a fatality, serious injury, or substantial property damages. The board is authorized, among other things, to make recommendations to EPA. In September 2011, the Chemical Safety Board released a report investigating three incidents involving fatality and injuries at oil and gas storage tanks located at well sites and surveyed an additional 23 such incidents that occurred between 1983 and 2010.[132] The report found that these accidents occurred when the victims—all young adults—gathered at rural unmanned oil and gas storage sites lacking fencing and warning signs. This report concluded such sites pose a public safety risk. The report also reviewed federal, state, and local regulations, inherently safer designs of tanks, and industry standards. Noting that exploration and production storage tanks are exempt from the security requirements of CWA[133] and from the risk management requirements of CAA,[134] the Chemical Safety Board recommended that EPA encourage owners and operators to reduce these risks.[135] Specifically, the Chemical Safety Board recommended EPA "publish a safety alert directed to owners and operators of exploration and production facilities with flammable storage tanks, advising them of their general duty clause responsibilities for accident prevention under CAA." The letter requests that EPA provide within 180 days a response stating how EPA will address the recommendation.[136] On June 27, 2012, EPA responded to the Chemical Safety Board and stated that EPA agrees to develop and publish a safety alert and anticipates the agency will be able to publish a final safety alert

[132]U.S. Chemical Safety and Hazard Investigation Board, *Investigative Study Final Report: Public Safety at Oil and Gas Storage Facilities*, Report No. 2011-H-1 (September 2011) [hereinafter Chemical Safety Board Report].

[133]Note that the Clean Water Act spill prevention control, and countermeasure (SPCC) rule requires oil and gas production facilities meeting applicability thresholds to prepare SPCC plans generally by November 2011, or before commencing operation. 40 C.F.R. pt. 112 (2012). However, oil production facilities are excluded from the SPCC regulations' security provisions. Id. at § 112.7(g).

[134]Chemical Safety Board Report at 8. See 40 C.F.R. § 112.7(g) (2012).

[135]Chemical Safety Board Report at 52.

[136]See CAA § 112(r)(6)(I), 42 U.S.C. § 7412(r)(6)(I) (2012). See also
http://www.csb.gov/recommendations/details.aspx?SID=95&pg=1&F_InvestigationId=95

by June 2013.[137] The Chemical Safety Board also made related
recommendations to several states and industry associations.[138]

EPA Enforcement Authorities

Even where a state implements key CAA provisions, EPA retains
oversight and enforcement authority. For example, EPA may initiate an
enforcement action via an administrative order or a civil action for a
violation of any requirement or prohibition of an applicable SIP, permit, or
certain other requirement or prohibition after notification to the state and
the party.[139] CAA also gives EPA authorities regarding access to records
and the ability to require provision of information, as to any person who
owns or operates any emission source, among others.[140]

Imminent and Substantial Endangerment Authority

Where EPA receives evidence that a source or a combination of sources
present an imminent and substantial endangerment to public health or
welfare, or the environment, EPA may bring suit or, where prompt action
is needed, issue orders to stop the emission of air pollutant or take other
necessary action.[141] EPA must first consult with state and local authorities
and attempt to confirm the accuracy of information before taking such
actions.

---

[137]Regarding the four specific items the Board recommended for inclusion, EPA stated
that it will address three, but the fourth item is under the jurisdiction of another agency.

[138]Chemical Safety Board Report at 52-54.

[139]CAA § 113(a)(1) and (3), 42 U.S.C. § 7413(a)(1) and (3) (2012); see generally CAA §
113, 42 U.S.C. § 7413 (2012).

[140]CAA § 114(a), 42 U.S.C. § 7414(a) (2012).

[141]CAA § 303, 42 U.S.C. § 7603 (2012).

Appendix V: Key Requirements and Authorities under the Resource Conservation and Recovery Act

In 1976, Congress passed the Resource Conservation and Recovery Act (RCRA), generally establishing EPA authority to regulate the generation, transportation, treatment, storage, and disposal of hazardous waste,[1] and also including some provisions respecting solid waste.[2] As to hazardous waste, EPA may authorize states to administer their own permitting programs in lieu of the federal program, as long as these state programs are equivalent to and consistent with the federal program and provide for adequate enforcement.[3] As to solid waste, RCRA provided a more limited federal role and included incentives for states to implement programs to manage nonhazardous solid waste disposal, a prohibition on open dumping of wastes, and a requirement for EPA to promulgate technical criteria for classifying solid waste disposal facilities, among other things.[4]

Subtitle C – Hazardous Waste

RCRA established federal requirements and EPA regulatory authority for "cradle-to-grave" management of hazardous wastes. RCRA defines hazardous waste as:

> a solid waste, or combination of solid wastes, which because of its quantity, concentration, or physical, chemical, or infectious characteristics may (A) cause, or significantly contribute to an increase in mortality or an increase in serious irreversible, or incapacitating reversible, illness; or (B) pose a substantial present or potential hazard to human health or the environment when improperly treated, stored, transported, or disposed of, or otherwise managed.[5]

EPA regulations implementing RCRA establish several means by which solid waste may be deemed hazardous for purposes of the Subtitle C regulations, including specifically being listed by EPA as a hazardous waste or by exhibiting one of the following four characteristics: toxicity,

---

[1]Pub. L. No. 94–580, 90 Stat. 2795 (1976) (codified as amended at 42 U.S.C. §§6901-6992k (2012)). Although RCRA amended the Solid Waste Disposal Act, Pub. L. No. 89–272, Title II, 79 Stat. 997 (1965), the amended law is nonetheless sometimes referred to as RCRA, a convention we follow here. Subtitle C of RCRA, 42 U.S.C. ch. 82, subch. III (§§ 6921-6939f), governs hazardous waste management. Hereinafter, references are to RCRA sections as amended.

[2]RCRA Subtitle D, 42 U.S.C. ch. 82, subch. IV (§§ 6941-6949a) (2012).

[3]RCRA § 3006(b), 42 U.S.C. § 6926(b) (2012).

[4]RCRA §§ 4006-09, 4004(a), 4005(a), 42 U.S.C. §§ 6946-49, 6944(a), 6945(a) (2012).

[5]RCRA § 1004(5), 42 U.S.C. § 6903(5) (2012).

ignitability, corrosivity, or reactivity.[6] The generation, transport, and disposal of wastes meeting the RCRA regulatory hazardous definition are generally subject to RCRA Subtitle C requirements, such as reporting, using a manifest, and disposing of the waste in approved ways, such as through hazardous waste landfill.[7]

Exemption of Certain Oil and Gas Production Wastes from Regulation as Hazardous Waste under RCRA Subtitle C

Notwithstanding the provisions for identifying hazardous wastes, the Solid Waste Disposal Act Amendments of 1980 created a separate process for certain oil and gas exploration and production wastes. Under the statute, these wastes would not be subject to regulation as hazardous waste under RCRA Subtitle C unless specific actions were taken.[8] The amendments required EPA to conduct and publish "a detailed and comprehensive study...on the adverse effects, if any, of drilling fluids, produced waters, and other wastes associated with the exploration, development, or production of crude oil or natural gas or geothermal energy on human health and the environment."[9] The study report was to "include appropriate findings and recommendations for Federal and non-Federal actions concerning such effects."[10]

The law further required EPA to either propose regulations for such wastes or determine that regulation was not warranted. Any such regulations would require congressional action to become effective:

> [T]he Administrator shall, after public hearings and opportunity for comment, determine either to promulgate regulations under this subchapter for drilling fluids, produced waters, and other wastes associated with the exploration, development, or production of crude oil or natural gas or geothermal energy or that such regulations are unwarranted. The Administrator shall transmit his decision, along with any regulations,

---

[6]40 C.F.R. §§ 261.3 (2012). Some wastes otherwise meeting the definition are excluded from regulation as a hazardous waste; see 40 C.F.R. § 261.4 (2012).

[7]See, e.g., RCRA §§ 3002, 3003, 3004, 42 U.S.C. §§ 6922, 6923, 6924 (2012).

[8]Pub. L. No. 96-482 § 7, 94 Stat. 2334, 2336 (1980), 42 U.S.C. § 6921(b)(2) (2012).

[9]Pub. L. No. 96-482 § 29(2), 94 Stat., 2350 (1980), amending RCRA § 8002(m), 42 U.S.C. § 6982(m) (2012).

[10]Id.

if necessary, to both Houses of Congress. Such regulations shall take effect only when authorized by Act of Congress.[11]

Pursuant to these provisions, EPA conducted the study and found that organic pollutants at levels of potential concern (levels that exceed 100 times EPA's health-based standards) included the hydrocarbons benzene and phenanthrene. Inorganic constituents at levels of potential concern included lead, arsenic, barium, antimony, fluoride, and uranium. EPA then issued a determination that regulation of oil and gas exploration and production wastes under RCRA Subtitle C was not warranted.[12] EPA focused on three key factors pertaining to these wastes: (1) the characteristics, management practices, and resulting impacts of these wastes on human health and the environment; (2) the adequacy of existing state and federal regulatory programs; and (3) the economic impacts of any additional regulatory controls on industry:[13]

In considering the first factor, EPA found that a wide variety of management practices are utilized for these wastes, and that many alternatives to these current practices are not feasible or applicable at individual sites...As to the second factor, EPA found that existing State and Federal regulations are generally adequate to control the management of oil and gas wastes. Certain regulatory gaps do exist, however, and enforcement of existing regulations in some States is inadequate. EPA's review of the third factor found that imposition of Subtitle C regulations for all oil and gas wastes could subject billions of barrels of waste to regulation under Subtitle C as hazardous wastes and would cause a severe economic impact on the industry and on oil and gas production in the U.S...and could cause severe short-term strains on the capacity of Subtitle C Treatment, Storage, and Disposal Facilities..and a significant increase in the Subtitle C permitting burden for State and Federal hazardous waste programs.[14]

---

[11]42 U.S.C. § 6921(b)(2)(B)-(C) (2012).

[12]53 Fed. Reg. 25,446, 25,447 (July 6, 1988). The study, titled "Management of Wastes from the Exploration, Development, and Production of Crude Oil, Natural Gas, and Geothermal Energy " was submitted to Congress in December 1987. See also Clarification of the Regulatory Determination for Wastes From the Exploration, Development and Production of Crude Oil, Natural Gas and Geothermal Energy, 58 Fed. Reg. 15,284 (Mar. 22, 1993).

[13]53 Fed. Reg. at 25,450.

[14]Id. at 25,446.

EPA stated that regulation of these wastes as hazardous waste under Subtitle C posed significant problems, including the lack of flexibility in the statute to take into account the varying geological, climatological, geographic, and other differences characteristic of oil and gas production sites, and to consider cost in applying the requirements—such that EPA would be unable to craft a program to avoid severe economic impacts and to fill only the gaps in existing programs.[15]

In lieu of regulating these wastes as hazardous waste under Subtitle C, EPA announced "a three-pronged approach toward filling the gaps in existing State and Federal regulatory programs," comprised of (1) improving existing programs under RCRA, the Safe Drinking Water Act, and the Clean Water Act; (2) working with states to improve their programs; and (3) working with Congress on any additional legislation that might be needed.[16] EPA further stated that it planned to revise its existing standards under Subtitle D of RCRA, "tailoring these standards to address the special problems posed by oil, gas, and geothermal wastes and filling the regulatory gaps,"[17] and "in developing these tailored Subtitle D standards for crude oil and natural gas wastes, EPA will focus on gaps in existing State and Federal regulations and develop appropriate standards that are protective of human health and the environment. Gaps in existing programs include adequate controls specific to associated wastes and certain management practices and facilities for large-volume wastes, including roadspreading, landspreading, and impoundments."[18]

As far as implementing the three-pronged approach, according to a 2011 EPA presentation, the agency developed Clean Water Act effluent guidelines for offshore and coastal oil and gas production, but EPA did not augment its RCRA Subtitle D regulations as planned, stating that it decided to work with the states instead.[19] EPA's work with states featured audits that ultimately led to the State Review of Oil and Natural Gas

---

[15]Id. at 25,447.

[16]Id.

[17]Id.

[18]Id. at 25,457.

[19]EPA, Exploration & Production Waste and RCRA, presented at ASTSWMO Annual Meeting (Oct. 26, 2011).

Environmental Regulations (STRONGER) program.[20] EPA also worked with industry representatives to develop best management practices for exploration and production wastes, but these efforts did not culminate in any document or guidance.

On September 8, 2010, the Natural Resources Defense Council submitted a petition requesting regulation of waste associated with the exploration, development, or production of oil, natural gas, and geothermal energy.[21] The petition asserts that EPA can and should revisit the determination not to regulate these wastes because, among other things, the underlying assumptions—concerning the availability of alternative disposal practices, the adequacy of state regulations, and economic harm to the oil and gas industry—are no longer valid. The petition requests that EPA promulgate regulations applying to wastes from the exploration, development and production of oil and natural gas under Subtitle C of RCRA.

EPA officials told us the petition is currently under consideration and that the agency has not established a time frame for its decision.[22] According to an EPA presentation, OSWER's Office of Resource Conservation and Recovery is currently (1) reviewing alleged incidents cited in the Natural Resources Defense Council petition; (2) compiling and reviewing state regulations in states with natural gas activities; and (3) reviewing best management practices for oil and gas exploration and production wastes developed by industry, federal, and state associations.[23] EPA does not anticipate releasing any studies, surveys, or other documents in the interim period. EPA officials said that when EPA is finished examining the

---

[20]The STRONGER program is conducted through the Ground Water Protection Council and brings together stakeholders to examine state oil and gas regulations and make recommendations for improvement.

[21]Letter, NRDC to EPA, Petition for Rulemaking Pursuant to Section 6974(a) of the Resource Conservation and Recovery Act Concerning the Regulation of Wastes Associated with the Exploration, Development, or Production of Crude Oil or Natural Gas or Geothermal Energy (Sept. 8, 2010); see also RCRA § 7004(a), 42 U.S.C. § 6974(a) (2012).

[22]See RCRA § 7004(a), 42 U.S.C. § 6974(a) (2012) (requiring only that EPA shall take action on a petition to promulgate regulations "within a reasonable time following receipt" of the petition.)

[23]EPA, Exploration & Production Waste and RCRA, presented at ASTSWMO Annual Meeting (Oct. 26, 2011).

issue, the agency intends to issue a proposed response to the petition. The proposed response will be printed in the *Federal Register*, and EPA will establish an electronic docket and provide an opportunity for public comment. Although EPA has not yet sought public comment on the petition, the agency has received several unsolicited comment letters, including from two industry associations, the STRONGER program, and two states.

If EPA revises the regulatory determination for some or all exploration and production wastes, the agency would conduct a full regulatory process to propose the regulations. Under the key RCRA provision, the regulations would not become effective until authorized by congressional action. Should the exemption be lifted, not all exploration and production wastes would necessarily be hazardous. Rather, whether particular exploration and production wastes would be hazardous and subject to regulation would depend on whether those particular wastes meet the regulatory definition of hazardous (i.e., are a listed waste or exhibit a characteristic of hazardous waste).

Oil and Gas Exploration and Production Wastes That Are Not Exempt from Regulation

While well sites wastes originating within the well or generated by field operations such as water separation, demulsifying, degassing, and storage are exempt, RCRA Subtitle C regulations generally apply to other wastes that may be generated at oil and gas wells, such as discarded unused products, solvents used to clean surface machinery, and others, if they are actually hazardous. In 2002, EPA published a guide titled "Exemption of Oil and Gas Exploration and Production Wastes from Federal Hazardous Waste Regulations" that identifies, among other things, a list of nonexempt wastes. The guide identified nonexempt wastes including the following wastes that may be generated by activities at oil and gas well sites:

- unused fracturing fluids or acids;

- painting wastes;

- waste solvents;

- oil and gas service company wastes such as empty drums, drum rinsate, sandblast media, painting wastes, spent solvents, spilled chemicals, and waste acids;

- vacuum truck and drum rinsate from trucks and drums transporting or containing nonexempt waste;

- used equipment lubricating oils;

- waste compressor oil, filters, and blowdown;

- used hydraulic fluids;

- caustic or acid cleaners;

- laboratory wastes;

- sanitary wastes;

- pesticide wastes;

- radioactive tracer wastes; and

- drums, insulation, and miscellaneous solids.

According to EPA's guidance document, this list represents some types of wastes that, if hazardous, are not exempt from Subtitle C regulation; however, these wastes may or may not be hazardous in a particular situation. These wastes are hazardous if they are a listed hazardous waste or exhibit a hazardous characteristic, such as ignitability or toxicity. If hazardous, then the facility is subject to waste management requirements that vary depending upon the amount of hazardous waste generated per calendar month.

RCRA regulations establish several categories for facilities generating hazardous waste, with differing reporting obligations.[24] Among these, the lowest level category is conditionally exempt small quantity generators, composed of facilities generating no more than 100 kilograms (220 pounds) per month of hazardous waste.[25] These facilities are subject to limits on the amount of hazardous waste they accumulate,[26] as well as

---

[24]40 C.F.R. § 261.5, pt. 262 (2012).

[25]Id. at § 261.5(a).

[26]Id. at (g)(2).

general requirements to determine which wastes are hazardous[27] and to ensure that any hazardous wastes sent for off-site disposal are sent to state-approved facilities, RCRA-permitted or interim status, or for certain wastes, universal waste facilities, facilities beneficially using, recycling, or reclaiming the waste.[28] Generally, conditionally exempt small quantity generators would not be required to have an EPA ID number.[29] Small quantity generators are those facilities generating more than 100 kilograms (220 pounds) but less than 1,000 kilograms (2,220 pounds) per month of hazardous waste. These facilities are subject to limits on the amount of hazardous waste they accumulate, as well as storage requirements,[30] and general requirements to determine which wastes are hazardous[31] and to ensure that any hazardous wastes sent for off-site disposal are sent to RCRA-permitted or interim status facilities.[32] In addition, the small quantity generators are required to have an EPA ID number and use manifests, by which hazardous waste may be tracked.[33]

For facilities, like oil and gas well sites, that may generate hazardous wastes but do not store, treat, or dispose of these wastes, no specific actions by EPA (or the authorized state) are required, beyond issuance of an EPA ID number to those facilities notifying EPA that it has generated hazardous waste in amounts making it a small or large quantity generator.[34] EPA uses the notification information and EPA ID number to identify the universe of regulated waste generators and their specific regulated waste activities, for tracking, and for a variety of enforcement

---

[27]Id. at § 262.11.

[28]Id. at § 261.5(g)(3). If the facility generates a subset of hazardous waste known as acute hazardous waste, it is subject to additional requirements for that waste. Id. at § 261.5(f)(3).

[29]Id. at §§ 261.5(b), 262.12.

[30]Id. at § 261.34.

[31]Id. at § 262.11.

[32]Id. at g)(3). If the facilities generate a subset of hazardous waste known as acute hazardous waste, it is subject to additional requirements for that waste. Id. at § 261.5(f)(3).

[33]Id. at §§ 262.12, 262.20.

[34]Id. at § 262.12(a) (a generator, other than a conditionally exempt small quantity generator, is essentially required to obtain an identification number before storing the waste or offering it to a transporter.).

and inspection purposes.[35] Generally, EPA (or the authorized state's) involvement at generator-only sites includes receiving notifications and issuing identification numbers, receiving biennial reports,[36] conducting compliance assurance activities such as inspections, and investigating alleged problems.

EPA has not undertaken a specific assessment of the extent to which oil and gas well sites are generating small amounts of regulated hazardous wastes and consequently are regulated as small quantity generators or conditionally exempt small quantity generators. EPA officials were unaware of the extent to which oil and gas well sites generate nonexempt hazardous waste (e.g., hazardous wastes other than exempt exploration and production wastes) in quantities significant enough to require an EPA ID number. EPA Region 8 officials were unaware of any instances in which a well site requested an EPA ID number. A challenge in understanding the extent to which oil and gas well sites are regulated stems in part from the use of North American Industry Classification System (NAICS) codes. While there is a code at the six-digit level that generally corresponds with oil and gas production,[37] it appears that, for some facilities with this code, the facility entry includes associated downstream facilities such as a compressor station or gas processing plant, making it impossible to use RCRAInfo – a publicly available EPA database that contains information on RCRA generators -- alone to identify well sites triggering the particular requirement of interest. For example, this database shows that some facilities with the oil and gas production NAICS code are listed as conditionally exempt small quantity generators. GAO's review of a small sample of these listings suggests

---

[35]See EPA Form 8700-12.

[36]Biennial reports are only required of large quantity generators. The reports are to include, among other things, a description of the generated hazardous wastes and the quantities shipped off-site to a U.S. treatment, storage, or disposal facility. 40 C.F.R. § 262.41 (2012).

[37]NAICS code 211111, Crude Petroleum and Natural Gas Extraction. A further complication in the context of this report is our scope is focused on unconventional resources produced using land-based wells, where the NAICS code at the six-digit level does not reflect these distinctions and includes other resources and offshore wells.

some may include downstream facilities, while others appear to be well sites.[38]

Subtitle D – Solid Waste

Oil and gas exploration and production wastes may be RCRA statutory solid wastes even if they are exempt from hazardous waste requirements or are nonhazardous wastes. As compared with hazardous waste, RCRA provided EPA a different and largely nonregulatory role for solid waste.[39] EPA's role in solid waste management is focused on assisting states in developing solid waste management programs. For example, EPA developed guidelines for certain aspects of solid waste management.[40] A key part of EPA's limited regulatory role[41] for solid waste was to establish criteria defining which solid waste disposal facilities and practices are "sanitary landfills" and those which constitute "open dumps,"[42] where RCRA prohibited open dumping of solid waste.[43]

[38] EPA officials from one Region believed some well sites may be small quantity generators of hazardous waste.

[39] See, e.g., RCRA §§ 1003(a)(1)-(3), 4001, 42 U.S.C. §§ 6902(a)(1)-(3), 6941 (2012).

[40] See, e.g., RCRA § 1008, 42 U.S.C. § 6907 (2012), 40 C.F.R. pt. 243 (2012) (Guidelines for the Storage and Collection of Residential, Commercial, and Institutional Solid Waste, containing guidelines that are recommended but not required for states).

[41] In addition, for two types of solid waste disposal facilities (those receiving conditionally exempt small quantity generator waste and household hazardous wastes), RCRA provided that states must implement an EPA-approved system to assure the facilities comply with EPA criteria, or the facilities would be subject to EPA hazardous waste enforcement authorities. RCRA § 4005(c)(1)(A)-(B), 42 U.S.C. § 6945(c)(1)(A)-(B) (2012) (providing that each State shall adopt and implement a permit program or other system of prior approval and conditions to assure that each solid waste management facility within such State which may receive hazardous household waste or hazardous waste [from small quantity generators] will comply with," respectively, the applicable initial and revised criteria established by EPA), RCRA § 4005(c)(1)(C), 42 U.S.C. § 6945(c)(1)(C) (2012) (EPA shall determine if each program is adequate); see 40 C.F.R. pts. 239, 257 subpt. B; 258 (2012).

[42] RCRA §§ 1008(a)(3), 4004, 42 U.S.C. §§ 6907(a)(3), 6944, 40 C.F.R. § 257.1, pt. 257, § 258.1(g), pt. 258 (2012). See also 53 Fed. Reg. 25,446 (July 6, 1988) (noting that "[t]he existing Federal standards under Subtitle D of RCRA provide general environmental performance standards for disposal of solid wastes, including oil, gas, and geothermal wastes, but these standards do not fully address the specific concerns posed by oil and gas wastes. Nevertheless, EPA has authority under Subtitle D to promulgate more tailored criteria.").

[43] RCRA § 4005(a), 42 U.S.C. §§ 6945(a) (2012).

Consistent with the scheme established by RCRA Subtitle D, states have primary responsibility for managing disposal of solid waste, including that resulting from oil and gas exploration and production. State solid waste programs regulate treatment (which may include incineration) and land disposal of these wastes, among other things. In addition, states may have specific programs to address oil and gas production wastes, and some states put such wastes in a special category of solid waste, such as industrial wastes, with more stringent requirements than the federal minimum requirements. (See report and app. IX for discussion of selected aspects of state waste management.)

Enforcement

EPA has certain enforcement authorities to address hazardous wastes. RCRA sections 3007, 3008, and 3013 collectively provide EPA with authorities to monitor compliance, conduct investigations, and enforce Subtitle C (the hazardous waste subtitle) and its implementing regulations.[44] Each of these key authorities depends, among other things, on the existence or presence of a hazardous waste in a given situation. EPA's authority under sections 3007 and 3013 extends beyond waste that is regulated as hazardous under Subtitle C (e.g., wastes meeting the regulatory definition of hazardous waste), and includes waste that meets the statutory definition of hazardous waste in RCRA section 1004(5).[45]

For example, section 3008(a) authorizes EPA to issue administrative compliance orders "whenever on the basis of any information" the EPA Administrator determines that any person has violated or is in violation of any requirement of Subtitle C.[46],[47] These orders may require the person to come into compliance immediately or by a specific time frame and/or

---

[44]RCRA §§ 3007, 3008, 3013, 42 U.S.C. §§ 6927, 6928, 6934 (2012).

[45]Cf. RCRA § 1004(5), 42 U.S.C. § 6903(5) (2012) with 40 C.F.R. §§ 261.3,-4 (2012).

[46]RCRA § 3008, 42 U.S.C. § 6928 (2012).

[47]EPA also has authorities to require corrective action or such other response measure as necessary to protect human health or the environment from past and present contamination, at RCRA permitted or interim status facilities—that is, where the facility is a storage, treatment, or disposal facility and has or should have a Subtitle C hazardous waste permit. EPA can use this corrective action authority only at facilities that are RCRA permitted or interim status facilities, and cannot require a corrective action at a generator-only facility. Interim status facilities are facilities that treat, store, or dispose of hazardous waste and have begun the process of applying for a RCRA permit. See RCRA §§ 3004(u)-(v), 3008(h), 42 U.S.C. §§ 6924(u)-(v), 6928(h) (2012).

pay a civil penalty for any past or current violation and may include suspension or revocation of a facility's RCRA permit. Alternatively, EPA, through the Department of Justice, may file a civil action in federal court for violations of RCRA and its implementing regulations and permits. EPA must give notice to the state, if it has an EPA-authorized hazardous waste program, prior to issuing an order or filing a civil judicial action.

Section 3007(a) gives EPA authority to inspect and copy records and to obtain samples from any person who generates, stores, treats, transports, disposes of, or otherwise handles or has handled hazardous wastes, and to enter sites where hazardous wastes are or have been generated, stored, treated, disposed of, or transported from.[48] Section 3007 also establishes mandatory compliance inspections.[49] EPA has interpreted its section 3007 authority, discussed above, to include the authority to access records and sites related to solid waste "that the Agency reasonably believes may pose a hazard when improperly managed."[50] EPA officials did not provide any examples of EPA using its section 3007 authority at oil or gas well sites.

Section 3013 authorizes EPA to issue an order requiring monitoring, testing, analysis, and reporting if the EPA Administrator determines, upon receipt of any information, that the presence or release of any hazardous waste at a facility or site at which hazardous waste is, or has been, stored, treated, or disposed of may present a substantial hazard to human health or the environment.[51] Furthermore, in certain circumstances, EPA may use its authority under section 3013 to conduct its own investigation into the nature and extent of a potential hazard.[52]

---

[48]RCRA § 3007(a), 42 U.S.C. § 6927(a) (2012).

[49]Id. at § 6927(e) (2012).

[50]See, e.g., 53 Fed. Reg. 25,446, 25,457 (1988) ("EPA believes this [section 3007] authority does not limit information collection to "hazardous" waste identified under Subtitle C, but also authorizes the collection of information on any solid waste that the Agency reasonably believes may pose a hazard when improperly managed. (EPA may also use this authority in preparing enforcement actions.)").

[51]RCRA § 3013(a), 42 U.S.C. § 6934(a) (2012).

[52]Id. at § 6934(d) (2012).

EPA officials did not provide any examples of EPA using these hazardous waste enforcement provisions for incidents arising at oil or gas well sites.

EPA has fewer enforcement responsibilities and authorities for nonhazardous waste facilities under RCRA Subtitle D, than it does for hazardous waste activities regulated under RCRA Subtitle C. In particular, state solid waste programs are based in state law and generally are not subject to enforcement or overfiling by EPA. RCRA's prohibition on open dumping of solid and hazardous waste is enforceable by citizen suit.[53]

Imminent and Substantial Endangerment Authority

EPA has imminent and substantial endangerment authority to address both hazardous and solid wastes. Section 7003 authorizes EPA to issue administrative orders and to file suit in federal district court. In addition, "upon receipt of evidence that the past or present handling, storage, treatment, transportation or disposal of any solid waste or hazardous waste may present an imminent and substantial endangerment to health or the environment," EPA has authority to restrain any person who has contributed or who is contributing to such handling, storage, treatment, transportation or disposal, from such activity, to order them to take such other action as may be necessary, or both.[54] Such orders can be issued to a person who contributed in the past or is currently contributing to the imminent and substantial endangerment to health or the environment.[55] Section 7003 orders are enforceable; if a nonfederal recipient fails to comply, EPA can enforce the order, including fines, by requesting that Department of Justice file suit in federal court.[56]

[53]RCRA § 7002, 42 U.S.C. § 6972 (2012). RCRA section 7002 authorizes citizen and state suits including against any person who is alleged to be in violation of any permit, standard, regulation, condition, requirement, prohibition, or order which has become effective pursuant to RCRA, or who has contributed or is contributing to the past or present handling, storage, treatment, transportation, or disposal of any solid or hazardous waste which may present an imminent and substantial endangerment to health or the environment. Citizen suits are subject to various conditions, such as a requirement for advance notice to EPA and the state, and that neither EPA nor the state is taking certain actions to address the problem.

[54]RCRA § 7003, 42 U.S.C. § 6973 (2012).

[55]Id. at § 6973(a).

[56]Id. at § 6973(b).

EPA's imminent and substantial endangerment authority is not limited to Subtitle C regulated hazardous wastes but also includes statutory solid wastes and hazardous wastes.[57] EPA has interpreted the authority broadly, to allow a range of actions to be taken, including addressing the threat of endangerment.[58] Nonetheless, EPA officials noted that a section 7003 action is distinct from, for example, the agency's Subtitle C enforcement authorities because the objective of such an action is to abate the imminent and substantial endangerment, rather than to enforce specific RCRA requirements. Whether RCRA section 7003 authority is applicable to a given situation requires a fact-based determination that the facts establish the statutory elements, including the existence of conditions that may present an imminent and substantial endangerment.

EPA has issued section 7003 orders at several facilities handling wastes from oil and gas well sites. For example, as previously discussed, EPA Region 8 participated in an effort with the FWS, states, and tribes, after the FWS expressed concerns about migratory birds landing on open pits that contained oil and water, which killed or harmed the birds.[59] The effort involved aerial surveys to observe pits. Where apparent problems were identified, relevant federal or state agencies were notified and were to give oil and gas operators an opportunity to correct problems. Ground inspections were then conducted where deemed warranted and, if problematic conditions were found, further follow up action was taken by EPA or the relevant state or other federal agency. As a result of this effort, EPA issued nine orders pursuant to RCRA section 7003 authority.[60] According to the report, the orders required operators "to remove oil from pits, install effective exclusionary devices, and/or clean up sites."[61] EPA Region 8 has issued section 7003 orders to several commercial oilfield waste disposal facility operators in Wyoming, finding

---

[57]RCRA § 1004(5), (27), 42 U.S.C. § 6903(5), (27) (2012).

[58]See EPA. Office of Enforcement and Compliance Assurance, Guidance on the Use of Section 7003 of RCRA (October 1997).

[59]EPA Region 8, Oil and Gas Environmental Assessment Effort 1996 – 2002, at v (2003).

[60]Id. at 8.

[61]Id.

each site endangered the environment including having caused bird mortalities due to inadequate pit management.[62]

As another example, in 2005, EPA Region 6 entered into an agreement with an exploration company and property owners at a site in Oklahoma where the contents of a well drilling waste pit had been relocated onto residential property; the agreement required the waste to be removed, among other things.[63]

[62]*In the Matter of Jim's Water Service*, Initial Administrative Order, EPA Docket No. RCRA-08-2011-0002 (June 23, 2011); *In the Matter of Pure Petroleum LLC*, Administrative Order, EPA Docket No. RCRA-08-2011-0003 (Sept. 16, 2011). See also Consent Decree, *United States of America v. High Plains Resources, Inc.*, No. 2:09-cv-00087 (Wy. Nov. 3, 2010).

[63]*West Bay Exploration*, Agreement and Remediation Plan, EPA Docket No. RCRA-06-2005-0913 (Sept. 14, 2005).

Appendix VI: Key Requirements and Authorities under the Comprehensive Environmental Response, Compensation, and Liability Act

In 1980, Congress passed the Comprehensive Environmental Response, Compensation, and Liability Act (CERCLA), often referred to as "Superfund," to address the cleanup of releases of hazardous substances, pollutants, and contaminants nationwide and, in so doing, protect human health and the environment from their effects.[1] The enactment of CERCLA gave the federal government the authority to respond to actual and threatened releases of hazardous substances, pollutants, and contaminants that may endanger public health or welfare or the environment,[2] as well as requiring reporting of hazardous substances releases above threshold quantities.[3] CERCLA also established a liability scheme, whereby potentially responsible parties such as owners and operators may be liable for cleanup and other costs stemming from the release (or threatened release) of hazardous substances into the environment from a facility.[4] CERCLA is primarily a remedial statute; it is preventive in that it authorizes responses to threatened releases of hazardous substances, pollutants, and contaminants, and to the extent that the liability scheme provides incentives for owners and operators to take care to avoid releases to the environment.

Relevant Exclusions and Definitions

Under a provision known as the petroleum exclusion, CERCLA's provisions do not apply to releases to the environment that are purely petroleum, including crude oil and natural gas, and fractions of crude oil including the hazardous substances, such as benzene, that are

---

[1]CERCLA, Pub. L. No. 96-510, 94 Stat. 2767 (1980) (codified as amended at 42 U.S.C. §§ 9601- 9675 (2012)). Hereinafter, references to CERCLA sections are as amended.

[2]CERCLA § 104, 42 U.S.C. § 9604 (2012).

[3]CERCLA § 103(a), 42 U.S.C. § 9603(a) (2012).

[4]Parties may also be held liable under CERCLA for damages related to the loss, injury or destruction of natural resources.

indigenous in those petroleum substances.[5] EPA can respond to releases of hazardous substances, however, even if there are colocated petroleum releases.[6]

CERCLA's liability and reporting provisions do not apply to federally permitted releases—generally, where a hazardous substance is released in compliance with a permit issued pursuant to certain federal environmental laws.[7] The statutory definition for such federally permitted releases exempt from CERCLA liability and reporting also includes:

> any injection of fluids or other materials authorized under applicable State law (i) for the purpose of stimulating or treating wells for the production of crude oil, natural gas, or water, (ii) for the purpose of secondary, tertiary, or other enhanced recovery of crude oil or natural gas, or (iii) which are brought to the surface in conjunction with the production of crude oil or natural gas and which are reinjected.[8]

However, EPA has explained, "[t]he National Response Center must be notified in any situation involving the use of injection fluids or materials that are not authorized specifically by State law for purposes of the development of crude oil or natural gas supplies and resulting in a release of a hazardous substance" at or above the threshold reporting quantity.[9]

---

[5]CERCLA § 101(14), 42 U.S.C. § 9601(14) (2012) (defining hazardous substance to exclude "petroleum, including crude oil or any fraction thereof which is not otherwise specifically listed or designated as a hazardous substance under [specified provisions of CWA, RCRA, CAA, and 15 U.S.C. § 2606], and the term does not include natural gas, natural gas liquids, liquefied natural gas, or synthetic gas usable for fuel (or mixtures of natural gas and such synthetic gas)."); CERCLA § 101(33), 42 U.S.C. § 9601(33) (with similar language, excluding petroleum from the definition of "pollutant or contaminant"). See also http://www.epa.gov/superfund/policy/release/rq/index.htm#exclude. Such releases may be reportable under provisions of other laws, such as the Oil Pollution Act of 1990 and Clean Water Act; see CWA § 311(b)(3)-(5), 33 U.S.C. § 1321(b)(3)-(5) (2012); 40 C.F.R. § 300.300(b) (2012).

[6]See http://www.epa.gov/superfund/policy/release/rq/index.htm#exclude. Releases of certain waste oils are also regulated under CERCLA. 40 C.F.R. § 302.4 (2012).

[7]CERCLA § 101(10), 42 U.S.C. § 9601(10) (2012). See also exclusions at id. § 9601(22) (2012).

[8]Id. at § 9601(10)(l) (2012). This provision was included in CERCLA as enacted in 1980. Pub. L. No. 96-510 § 101, 94 Stat. 2768 (1980). See also 53 Fed. Reg. 27,268 (July 19, 1988).

[9]53 Fed. Reg. at 27,275.

CERCLA Hazardous Substance Release Reporting

Where there has been a release of a hazardous substance, CERCLA section 103 requires a person in charge of a facility to report such releases above reportable quantities as soon as he/she has knowledge of such release to the National Response Center.[10] EPA regulations establish CERCLA hazardous substances and their reportable quantities.[11] While releases of pure petroleum (e.g., petroleum in which hazardous substances have not increased such as by addition or processing) are excluded, releases of CERCLA hazardous substances that are commingled with petroleum are subject to the reporting requirement.[12] Oil and gas well operators would be required to report any releases to the environment of other hazardous substances, for example, if a stored hazardous substance was accidentally spilled onto the ground, or if hazardous substances above the reportable quantity were injected but not authorized by state law.

The National Response Center—managed by the U.S. Coast Guard— receives release reports and forwards them to EPA Regions. When receiving a report, according to EPA Regional staff will screen the report for such factors as what was spilled and in what quantity and whether the spill threatens surface waters, to determine if EPA needs to respond and, if appropriate, will obtain additional information on the event, and/or send an on-scene coordinator to the site. EPA officials also noted they use the release reports to refer sites to program enforcement offices, such as the Clean Water Act's SPCC program, for follow-up. Although release reports are publicly available, the available search terms do not readily differentiate oil and gas well sites from other types of oil and gas facilities. EPA officials noted that there had been approximately 200 reports of oil spills from oil facilities in the last 5 years. EPA Region 5 officials stated that oil spills are more often related to pipelines, tank sites, or trucking accidents, with few occurring at well sites.

---

[10]CERCLA § 103(a), 42 U.S.C. § 9603(a) (2012). The National Response Center is the sole federal point of contact for reporting all hazardous substances and oil spills that trigger federal notification requirements under several laws. Information reported to the Center is disseminated to other agencies, such as EPA, as well as to states.

[11]40 C.F.R. pt. 302, § 302.4 at table 302.4 (2012).

[12]See http://www.epa.gov/superfund/policy/release/rq/index.htm#exclude. Releases of certain waste oils are also regulated under CERCLA. 40 C.F.R. § 302.4 (2012).

Relevant EPA Authorities

EPA established the Superfund program to carry out its responsibilities and authorities under CERCLA.[13] Under the Superfund program, EPA implements its authorities to compel parties responsible[14] for contaminating sites—via releases of hazardous substances—to clean them up, as well as to enter into agreements with such parties for them to conduct the cleanup. In addition, EPA can itself conduct response actions, which may include investigations and cleanup activities, and then seek reimbursement from the responsible parties.

The Superfund cleanup process involves a series of steps during which specific activities—such as investigations and cleanups—take place or decisions are made. The CERCLA program has two basic types of cleanup: (1) cleanups under the removal process, which generally address short-term threats, and (2) cleanups under the remedial action process, which are generally longer-term cleanup actions.[15] In determining whether to use removal or remedial authority to take a response action, EPA considers the time-sensitivity, complexity, comprehensiveness, and cost of the response action.[16]

Several EPA Superfund authorities are particularly relevant to oil and gas well operations, including the following:

- *Investigations, monitoring, coordination.* Under section 104(b), EPA generally may conduct investigation activities with appropriated program funds whenever a hazardous substance is released or there is a substantial threat of such a release, or there is reason to believe a

---

[13]Through Executive Order 12580, Superfund Implementation (1987), EPA was delegated key regulatory and enforcement authorities CERCLA granted to the President. In addition, CERCLA, as amended, granted certain authorities directly to the EPA Administrator.

[14]Under CERCLA, potentially responsible parties generally include current or former owners and operators of a site or the generators or transporters of the hazardous substances.

[15]40 C.F.R. § 300.5 (2012) (defining removal as including containment and removal of hazardous substances or other actions as may be necessary to minimize or mitigate damage to the public health or welfare of the United States or to the environment, and defining remedial action as including actions consistent with permanent remedy to prevent or minimize the release so that they do not migrate to cause substantial danger to present or future public health or welfare or the environment.) For more information, see 40 C.F.R. § 300.415 (removals), 300.430, 300.435 (remedial actions).

[16]EPA, Memorandum, "Use of Non-Time Critical Removal Authority in Superfund Response Actions" (2000).

release has occurred or is about to occur.[17] These activities may include monitoring, surveys, testing, and other information gathering, as well as planning, legal, fiscal, economic, engineering, architectural, and other studies or investigations, as deemed appropriate.[18]

- *Information gathering and access.* Under section 104(e), EPA has authority to obtain information as well as authorities to enter property and to conduct inspections and take samples.[19] Specifically, EPA may require a person to furnish information about the identification, nature, and quantity of materials that have been or are generated, treated, stored, or disposed of at a facility or transported thereto, or the nature or extent of a release or threatened release of a hazardous substance or pollutant or contaminant, or the ability of a person to pay for or to perform a cleanup, including related documents and records, among other things.[20] Where there is a reasonable basis to believe there may be a release or threat of release of a hazardous substance or pollutant or contaminant, EPA is authorized to enter a facility or property where such release is or may be threatened, among other things, and may inspect and obtain samples.[21] EPA may obtain access by agreement, warrant, or administrative order.[22] If consent is not granted, EPA may issue administrative orders or, through the Department of Justice, file civil actions, to compel compliance with requests made under these provisions.[23]

- *Removals.* Under section 104(a), EPA generally has authority to act whenever there has been a release or substantial threat of release into the environment of any hazardous substance. EPA generally may

---

[17]CERCLA § 104(a)(1), (b), 42 U.S.C. § 9604(a)(1), (b) (2012).

[18]Id.

[19]Id. at § 9604(e).

[20]Id. at § 9604(e)(1)-(2).

[21]Id. at § 9604(e)(3)-(4).

[22]Id. at § 9604(e)(4)-(5).

[23]Id. at § 9604(e)(5).

conduct removal actions, among other things.[24] Removal actions are broadly defined and include actions to monitor, assess, and evaluate the release; the disposal of removed material; and other actions to prevent, minimize, or mitigate damage to the public health or welfare or to the environment such as provision of alternative drinking water supplies.[25]

- *Imminent and substantial endangerment authority related to releases of a pollutant or contaminant.* Under section 104(a), EPA has authority to act whenever a release or substantial threat of release into the environment of any pollutant or contaminant may present an imminent and substantial danger to the public health or welfare. This provides EPA with authority over releases of substances that are not CERCLA hazardous but that may harm public health or welfare;[26] however, as noted above, releases that are purely petroleum are excluded. Under this authority, EPA may conduct removals, provide for remedial action, or take any other response measure consistent with the National Contingency Plan.[27]

- *Authorities to pursue potentially responsible parties.* In addition, under section 106(a), EPA, through the Department of Justice, can pursue injunctive relief in court, where an actual or threatened release of a hazardous substance from a facility may pose an imminent and substantial endangerment to the public health or welfare or the environment.[28] EPA also can issue an administrative order requiring a potentially responsible party to take response

---

[24]Id. at § 9604(a). In addition to removal actions, EPA may also conduct remedial actions at nonfederal sites which are listed on the National Priorities List, but it is somewhat unlikely an oil and gas well site would be listed on the National Priorities List in light of the petroleum exclusion, among other factors. CERCLA §§ 104(a), (c)(1), 111(e), 42 U.S.C. §§ 9604(a), (c)(1), 9611(e) (2012). The National Priorities List includes sites that EPA determines are among the nation's most seriously contaminated hazardous waste sites to receive attention under the federal Superfund program.

[25]CERCLA § 101(23), 42 U.S.C. § 9601(23) (2012).

[26]See id. at § 9601(33) . EPA cannot, however, recover its response costs associated with these releases.

[27]EPA has promulgated regulations comprising the National Oil and Hazardous Substances Pollution Contingency Plan. 40 C.F.R. pt. 300 (2012). This plan outlines procedures and standards for implementing the Superfund program.

[28]CERCLA § 106(a), 42 U.S.C. § 9606(a) (2012).

actions as may be necessary to protect public health and welfare and the environment.[29] CERCLA also provides authorities for EPA to pursue cleanup and related costs from potentially responsible parties, and to enter settlements, as well as providing for liability of potentially responsible parties for damages to federal, state, and tribal natural resources.[30]

EPA has utilized its CERCLA authorities at several locations where it has been alleged that hazardous substance releases from oil and gas well sites have contaminated land or groundwater. In an example at a conventional oil well, in the 1990s, EPA, as represented by the Department of Justice, reached an agreement in which an oil exploration and production company pled guilty to a criminal felony count related to CERCLA violations when operators disposed of waste oil and hazardous substances by injecting them down the annuli (the space between the well casing and the surrounding rock) of the oil wells, over a 2-year period.[31] According to the Department of Justice, the company agreed to spend $22 million to resolve the criminal case and related civil claims, which included claims brought under RCRA, SDWA, and EPCRA, as well as CERCLA.[32]

More recently, EPA has used CERCLA authorities to conduct response activities, investigations, and to obtain records relating to alleged hazardous substance or pollutant or contaminant releases from oil and gas well sites. For example, EPA used CERCLA section 104(a) to undertake emergency removal actions including well sampling and provision of alternate water supplies at a site in Dimock, Pennsylvania.[33] EPA is using CERCLA section 104(b) authority to conduct groundwater

---

[29]Id. at § 9606(a).

[30]See, e.g., CERCLA §§ 107, 122, 42 U.S.C. §§ 9607, 9622 (2012).

[31]See DOJ, news release (Sept. 23, 1999).

[32]See DOJ, news release (Sept. 23, 1999). See also, *United States of America v. BP Exploration (Alaska) Inc.*, Plea Agreement (Sept. 23, 1999), Stipulation of Settlement and Order (Sept. 23, 1999), Docket no. 3:99-cv-00549-JKS (D. Alaska).

[33]See Richard M. Fetzer, On-Scene Coordinator EPA, Action Memorandum to Dennis Carney, Associate Division Director, Hazardous Site Cleanup Division, EPA, re: Request for Funding for a Removal Action at the Dimock Residential Groundwater Site, Jan. 19, 2012.

contamination investigations at Pavillion, Wyoming.[34] EPA officials also referenced CERCLA section 104(e) authority in requesting information from operators of wells proximate to the Pavillion site.

EPA has used CERCLA section 104(e) in conjunction with other authorities in several "multimedia" information requests, where EPA seeks information under multiple statutes and for multiple media—air, land, water—that may be affected. In 2011, for example, EPA used CERCLA and other authorities to request information concerning a blowout at a Marcellus shale natural gas well in Bradford, Pennsylvania. In this instance, a well blowout during hydraulic fracturing resulted in the release of flowback fluids to a tributary of the Susquehanna River, as well as combustible gases to the atmosphere.[35]

---

[34]See EPA, Draft Report, *Investigation of Ground Water Contamination near Pavillion, Wyoming*, EPA 600/R-00/000 (December 2011) at xi, 1.

[35]Agency for Toxic Substances and Disease Registry, Health Consultation, Chesapeake ATGAS 2H Well Site, Leroy Hill Road, Leroy Township, Bradford County, PA (2011).

Appendix VII: Key Requirements and Authorities under the Emergency Planning and Community Right-to-Know Act

The Emergency Planning and Community Right-to-Know Act of 1986 (EPCRA) provides a mechanism to help communities plan for emergencies involving extremely hazardous substances, and to provide individuals and communities with access to information regarding the storage and releases of certain toxic chemicals, extremely hazardous substances, and hazardous chemicals in their communities.[1]

Generally Applicable Chemical Information, Inventory, and Release Reporting

EPCRA imposes a set of generally applicable requirements to report information on the uses, inventories, and releases into the environment of hazardous and toxic chemicals above threshold quantities.[2] Regarding releases, EPCRA section 304 requires owners or operators of facilities where a chemical is produced, used, or stored to notify state and local emergency planning authorities of certain releases.[3] The releases for which EPCRA requires reporting partially overlap with those for which the Comprehensive Environmental Response, Compensation, and Liability Act of 1980 (CERCLA)[4] requires reporting.[5] Where there is overlap, EPCRA's procedures ensure state and local authorities receive this

---

[1]Pub. L. No. 99–499, Title III, 100 Stat. 1728 (1986) (codified at 42 U.S.C. ch. 116 (2012)). Hereinafter, references to EPCRA sections are as amended.

[2]In addition to EPCRA sections 304, 311, and 312, 42 U.S.C. §§ 11004, 11021, 11022 (2012), provisions discussed herein, section 302, 42 U.S.C. § 11002 (2012) requires the owner or operator of a facility to provide notification to the state and local emergency planning authorities with jurisdiction over the facility within 60 days if any extremely hazardous substances—including ammonia, hydrofluoric acid, and others—are present at or above its threshold planning quantity. See generally 40 C.F.R. pt. 355 (2012).

[3] EPCRA § 304(a), 42 U.S.C. §§ 11004(a) (2012). See also EPA, List of Lists (2011) available at http://www.epa.gov/emergencies/docs/chem/list_of_lists_revised_7_26_2011.pdf The reporting requirement does not apply, however, to releases which results in exposure to persons solely at the site where the facility is located, nor to federally permitted releases under the CERCLA definition. Id. at § 11004(a)(2), (4).

[4]Pub. L. No. 96–510, 94 Stat. 2767 (1980) (codified at 42 U.S.C. ch. 103 (2012)).

[5]See 40 C.F.R. § 355.60 (2012). Three types of releases must be reported: (1) release of extremely hazardous substances for which notification is also required under CERCLA § 103(c), (2) release of extremely hazardous substances for which notification is not required under CERCLA § 103(c), but above reporting thresholds and subject to additional conditions, and (3) release of other hazardous substances for which notification is also required under CERCLA § 103(c), subject to CERCLA reporting thresholds or 1 pound default threshold.

information, and CERCLA's procedures ensure federal authorities receive notification.

Regarding reporting of chemical information and inventories, EPCRA sections 311 and 312 requirements apply only to those facilities storing or using (1) more than 500 pounds or the threshold planning quantity, whichever is lower, of extremely hazardous substances, or (2) more than 10,000 pounds of other hazardous chemicals.[6] These facilities are required to provide chemical information (e.g., Material Safety Data Sheet or other detailed list) and submit an annual inventory report to state and local emergency planning authorities and to the local fire department with jurisdiction over the facilities.[7]

Requirements under EPCRA That May Be Triggered at Well Sites

Well sites are subject to EPCRA sections 304, 311, and 312, among others, and may be subject to reporting requirements to the extent that the chemicals used, stored, or produced at well sites meet the respective reporting thresholds. Under EPCRA section 304, any facility, such as a well site, that produces, uses, or stores any hazardous chemical and has a release above the reportable quantity of a CERCLA hazardous substance or an extremely hazardous substance, must provide notification to state and local emergency planning authorities, as well as the National Response Center.[8] Under EPCRA sections 311 and 312, any facility, such as a well site, at which an extremely hazardous chemical or any other hazardous chemical is present at the relevant threshold quantity, must meet inventory reporting requirements. For extremely hazardous chemicals, the threshold is 500 pounds or its threshold planning quantity, whichever is less. For all other hazardous chemicals, the reporting threshold is 10,000 pounds.[9] For example, if the

---

[6]EPCRA §§ 302(b)(1), 304(a), 42 U.S.C. §§ 11002(b)(1), 11004(a) (2012); 40 C.F.R. § 370.10(a) (2012). Hazardous chemicals are defined as any chemical which is a physical hazard or a health hazard. EPCRA § 311(e), 42 U.S.C. § 11021(e) (2012), 40 C.F.R. § 355.61, 370.66 (2012), 29 C.F.R. § 1910.1200(c) (2012).

[7]EPCRA § 302(b)(1), 311(a)-(b), 312(a)-(b), 42 U.S.C. §§ 11002(b)(1), 11021, 11022 (2012), and 40 C.F.R. § 370.44 (2012).

[8]The reporting trigger for Section 304—that a facility produce, use, or store any hazardous chemical—does not require that the chemicals be present or stored on the site for any minimum period of time.

[9]40 C.F.R. § 370.10(a)(2)(i) (2012).

GAO-12-874 Unconventional Oil and Gas Development

aggregate amount of hydrofluoric acid, an extremely hazardous chemical with a threshold planning quantity of 100 pounds, at a well site exceeds that threshold then the facility must report under sections 311 and 312. As another example, if a well stores or uses more than 10,000 pounds of drip gas or natural gas condensate at any one time, then the facility must report under sections 311 and 312.

The extent to which these requirements are triggered at oil and gas well sites depends on the presence and quantities of listed chemicals at such sites, among other things. We did not locate any publicly available data on the quantity of chemicals stored at actual or typical well sites, but FracFocus[10] provides self-reported data on the types of chemicals used in hydraulic fracturing, meaning that these chemicals are present and used at well sites. According to data in FracFocus, some hydraulic fracturing operations may use various hazardous chemicals, including some that are also CERCLA hazardous substances, such as hydrochloric acid, formaldehyde, formic acid, acetaldehyde, ethylene glycol, methanol, acetic acid, sodium hydroxide, potassium hydroxide, acrylamide, and naphthalene; of these, one is also considered "extremely hazardous."[11]

According to EPA, its Regional offices have several cases in development where the facility triggered the reporting requirements under 311 and 312 during all phases of operation, including drilling, hydraulic fracturing, and production.[12] EPA stated that, based on the Regions' experience, section 311 and 312 requirements could be triggered at every well site. EPA provided an example of section 312 information for a well site, which according to EPA officials, indicates that some hazardous chemicals may

---

[10]FracFocus is the national hydraulic fracturing chemical registry managed by the Ground Water Protection Council and Interstate Oil and Gas Compact Commission. The Ground Water Protection Council is a national association of state groundwater and underground injection control agencies whose mission is to promote the protection and conservation of groundwater resources for all beneficial uses, recognizing groundwater as a critical component of the ecosystem. The Interstate Oil and Gas Compact Commission is a multistate government agency that promotes the conservation and efficient recovery of domestic oil and natural gas resources while protecting health, safety, and the environment.

[11]Cf. http://fracfocus.org/chemical-use/what-chemicals-are-used and 40 C.F.R. § 302.4 (2012).

[12]EPCRA sections 311 and 312, 42 U.S.C. §§ 11021, 11022 (2012), require reporting to the state and local emergency planning authorities and to the local fire department with jurisdiction over the facility; EPA does not receive these reports.

be present at the particular well site in quantities that would trigger section 311 and 312 requirements.[13]

The information provided by EPA suggests that the types of chemicals with maximum on-site quantities of 10,000 to 99,999 pounds are the following:

- cement and associated additives;

- silica;

- shale control additives;

- drilling mud and associated additives;

- deflocculants;

- lubricants, drilling mud additives; and

- alkalinity and pH control material.

The information provided by EPA also suggests that the types of chemicals with maximum on-site quantities of 100,000 to 999,999 pounds are the following:

- produced hydrocarbons,

- salt solutions,

- weight materials, and

- fuels.

Toxic Release Inventory

EPCRA also requires some facilities in listed industries to report to EPA their releases of listed toxic chemicals to the environment;[14] at present, these requirements do not apply to oil and gas well operations.

---

[13]EPA also provided information from an example of a section 312 form from a field service provider for the provider's facility where larger quantities of chemicals are stored and then loaded on trucks to service the wells.

Section 313 of EPCRA generally requires certain facilities that manufacture, process, or otherwise use any of more than 600 listed individual chemicals and chemical categories, to report annually to EPA and their respective state, for those chemicals used above threshold quantities. Facilities need to report the amounts that they released to the environment and whether they were released into the air, water, or soil.[15]

EPCRA further requires EPA to make this information available to the public,[16] which the agency does electronically through the Toxics Release Inventory (TRI) database. The Pollution Prevention Act of 1990 requires covered facilities that report to the TRI to also provide certain information about their waste management practices, including amounts of covered chemicals recycled or treated.[17] The purposes of making this information available include to inform citizens about releases of toxic chemicals to the environment; to assist governmental agencies, researchers, and other persons in the conduct of research and data gathering; and to aid in the development of appropriate regulations, guidelines, and standards, and for other similar purposes.[18]

EPCRA section 313(b)(1) specifies that these requirements shall apply to owners and operators of facilities meeting three conditions: (1) having 10 or more full-time employees;[19] (2) in certain Standard Industrial Classification codes; and (3) that manufactured, processed, or otherwise used a listed toxic chemical in excess of the reporting threshold during the calendar year. The law specified the Standard Industrial Classification codes subject to the reporting requirement. EPA has, from time to time, amended its regulations to reflect industry codes in use; first, providing a crosswalk from Standard Industrial Classification codes to the North

---

[14]The Pollution Prevention Act of 1990, Pub. L. No. 101–508 , Title VI, 104 Stat. 1388–321 (1990) (codified at 42 U.S.C. ch. 133 (2012)) requires facilities subject to EPCRA section 313 to also report annually toxic chemical source reduction and recycling activities. 42 U.S.C. § 13106(a)-(b) (2012).

[15]EPCRA § 313(g)(1)(C), 42 U.S.C. § 11023(g)(1)(C) (2012).

[16]Id. at (i).

[17]Pollution Prevention Act, § 6607, 42 U.S.C. § 13106 (2012).

[18]EPCRA § 313(h), 42 U.S.C. § 11023(h) (2012).

[19]Full-time employee is defined as 2,000 hours per year of full-time equivalent employment. A facility would calculate the number of full-time employees by totaling the hours worked during the calendar year by all employees, including contract employees, and dividing that total by 2,000 hours. 40 C.F.R.§ 372.3 (2012).

American Industry Classification System (NAICS) codes, and subsequently to update as needed to reflect changes to the NAICS codes.[20]

EPCRA section 313(b)(1)(B) provides EPA with authority to add or delete industrial codes.[21] EPA issued initial regulations to implement the TRI in 1988.[22] In the initial regulations, EPA discussed its approach to evaluating additional industrial codes under its discretionary authority but did not add any at that time.[23] Oil and gas extraction industries were not included on the statutory list of Standard Industrial Classification codes and hence were not subject to the rule.

EPA has since expanded the list of covered industries, but it has not included oil and gas extraction.[24] According to EPA's Sector Notebook, the addition of the oil and gas extraction industry to regulation under EPCRA section 313 has been a long-term consideration.[25] In 1997, pursuant to section 313(b)(1)(B), EPA added seven industry groups[26] to the list of industries required to report releases in a rulemaking known as the Industry

---

[20]Community Right-to-Know; Toxic Chemical Release Reporting Using NAICS, 71 Fed. Reg. 32,464, 32,465 (June 6, 2006); see generally http://www.epa.gov/tri/lawsandregs/naic/ncodes.htm

[21]EPCRA § 313(b)(1)(B), 42 U.S.C. § 11023(b)(1)(B) (2012). EPA can also add individual facilities. EPCRA § 313(b)(2), 42 U.S.C. § 11023(b)(2) (2012).

[22]53 Fed. Reg. 4500 (Feb. 16, 1988).

[23]Id. at 4503 (stating "EPA has discretionary authority to modify the coverage of facilities under section 313(b)(1)(B). The report of the congressional conference committee for Title III states that any such modifications are limited "* * * to adding [Standard Industrial Classification] codes for facilities which, like facilities within the manufacturing sectors Standard Industrial Classification codes 20 through 39, manufacture, process or use toxic chemicals in a manner such that reporting by these facilities is relevant to the purposes of this section... The Agency is choosing not to modify the facility coverage of the rule at this time...The Agency must carefully evaluate additional types of facilities that may be manufacturing, processing, or using listed toxic chemicals as well as facilities in [Standard Industrial Classification] codes 20 through 39 that do not handle such chemicals.").

[24]See http://www.epa.gov/tri/lawsandregs/naic/ncodes.htm.

[25]EPA Office of Compliance, Sector Notebook Project: Profile of the Oil and Gas Extraction Industry 114 (October 2000).

[26]These industries included metal mining, coal mining, electrical utilities that combust coal and/or oil for the purpose of generating power for distribution in commerce, certain hazardous waste processing or destruction facilities regulated by EPA, chemical wholesalers, petroleum terminals and bulk stations and solvent recovery services.

Expansion Rule.[27] Oil and gas exploration and production was among nine candidate industries considered in EPA's screening process. The preamble to the proposed Industry Expansion Rule stated in part:

> One industry group, oil and gas extraction classified in [Standard Industrial Classification] code 13, is believed to conduct significant management activities that involve EPCRA section 313 chemicals. EPA is deferring action to add this industry group at this time because of questions regarding how particular facilities should be identified. This industry group is unique in that it may have related activities located over significantly large geographic areas.

> While together these activities may involve the management of significant quantities of EPCRA section 313 chemicals in addition to requiring significant employee involvement, taken at the smallest unit (individual well), neither the employee nor the chemical thresholds are likely to be met. EPA will be addressing these issues in the future.[28]

The preamble of the final rule stated in part, "[a] number of commenters support EPA's decision not to include oil and gas exploration and production in its proposal, and urge EPA not to propose adding this industry in the future. EPA considered the inclusion of this industry group prior to its proposal, and indicated in the proposal that one consideration for not including it was concern over how a 'facility' would be defined for purposes of reporting in EPCRA section 313 …This issue, in addition to other questions, led EPA to not include this industry group. EPA will continue its dialogue with the oil and gas exploration and production industry and other interested parties, and may consider action on this industry group in the future."[29]

---

[27]Addition of Facilities in Certain Industry Sectors; Revised Interpretation of Otherwise Use; Toxic Release Inventory Reporting; Community Right-to-Know, 62 Fed. Reg. 23,834 (May 1, 1997) (known as the Industry Expansion Rule). In announcing the rule, EPA stated, "EPA believes that [section 313(b)(1)(B)] grants the Agency broad, but not unlimited, discretion to add industry groups to the facilities subject to EPCRA section 313 reporting requirements where EPA finds that reporting by these industries would be relevant to the purposes of EPCRA section 313." See also 71 Fed. Reg. 32,464, 32,465 (June 6, 2006).

[28]61 Fed. Reg. 33,588, 33,592 (June 27,1996).

[29]62 Fed. Reg. 23,834, 23,855 (1997).

In fall 2011, EPA conducted a discussion forum on regulations.gov. The background information provided in the forum stated that EPA was considering a rule to add or expand coverage to the following industry sectors: Iron Ore Mining, Phosphate Mining, Solid Waste Combustors and Incinerators, Large Dry Cleaners, Petroleum Bulk Storage, and Steam Generation from Coal and/or Oil.[30] EPA officials told us that, for the current possible rulemaking, the initial screening process for sectors to consider adding to the TRI included review of those sectors, such as oil and gas production, that were considered but ultimately not added in the 1997 rule. In addition, EPA officials said the initial screening process also included sectors covered by analogous registries of other countries. According to EPA, the oil and gas sector falls into both categories and was considered in the initial screening. As of July 2012, EPA officials stated that EPA does not anticipate adding oil and gas exploration and production sites as part of the possible rule currently under consideration to add industry sectors to the scope of TRI.[31] EPA officials explained that the agency has not changed its assessment of the oil and gas sector as it pertains to TRI reporting since the 1996 proposed rule and stated that adding oil and gas well sites would likely provide a substantially incomplete picture of the chemical uses and releases at these sites, and would therefore be of limited utility in providing information to communities.

EPCRA section 313 also specified the chemicals subject to the reporting requirement and provided a process and criteria for EPA to add or delete chemicals from the list.[32] In the proposal to the 1997 Industry Expansion Rule discussed above, EPA stated that oil and gas extraction activities "may involve the management of significant quantities of EPCRA section 313 chemicals."[33] In response to our request for background data regarding these chemicals and their quantities, EPA officials said they were unable to locate any record of the specific chemicals referred to in the 1996 proposal as being managed in "significant quantities." However,

---

[30] http://exchange.regulations.gov/exchange/topic/trisectorsrule/agencyintro/tri-exchange

[31] In July 2012, EPA officials stated that if the sector is not proposed to be added to the TRI, the agency does not anticipate placing documents related to EPA's consideration of the oil and gas production sector in the docket for any possible TRI rulemaking.

[32] See, e.g., 75 Fed. Reg. 72,727 (Nov. 6, 2010) (EPA's most recent addition to the TRI list of chemicals, adding 16 chemicals reasonably expected to be carcinogenic).

[33] 61 Fed. Reg. 33,588, 33,592 (June 27,1996).

EPA officials noted that Canada's National Pollutant Release Inventory (NPRI) has data on Canadian oil and gas wells for some TRI chemicals. Specifically, EPA identified several TRI chemicals that were also reported to the Canadian NPRI by oil and gas facilities as being released, disposed of, and/or transferred in large quantities in reporting year 2010 in Canada including ammonia, arsenic, cadmium, copper, hexavalent chromium, hydrogen sulfide, lead, manganese, mercury, phenanthrene, phosphorus, sulfuric acid aerosols, and zinc compounds.

If oil and gas exploration and production were added to the industries required to report to the TRI, such facilities meeting relevant thresholds would have to report releases of hydrogen sulfide, which is among the chemicals of particular concern some have cited. In October 2011, EPA lifted its administrative stay of the EPCRA section 313 reporting requirements for hydrogen sulfide, which had been in effect since 1994, shortly after the chemical was added to the list of toxic chemicals.[34] EPA conducted a technical evaluation of hydrogen sulfide and found no basis for continuing the administrative stay of the reporting requirements. The first reports under EPCRA section 313 for hydrogen sulfide will be due on July 1, 2013, for reporting year 2012.[35]

Enforcement

EPCRA provides EPA with various authorities to enforce the act's requirements.[36] For example, for violations of EPCRA section 311 or section 312 requirements, such as provision of annual inventory reports to state and local authorities, EPA may assess administrative penalties, or initiate court actions to assess civil penalties.[37] In cases of violations of section 304 release reporting requirements, EPA may assess administrative penalties, among other things.[38]

---

[34]76 Fed. Reg. 64,022 (Oct. 17, 2011).

[35]Id. at 64,025.

[36]EPCRA § 325, 42 U.S.C. § 11046 (2012).

[37]Id. at § 11046(c)(1), (2), (4).

[38] Id. at § 11046(b).

Appendix VIII: Key Requirements and Authorities under the Toxic Substances Control Act

To help protect human health and the environment, the Toxic Substances Control Act (TSCA) authorizes EPA to regulate the manufacture, processing, use, distribution in commerce, and disposal of chemical substances and mixtures.[1] EPA has authorities by which it may assess and manage chemical risks, including (1) to collect information about chemical substances and mixtures; (2) upon making certain findings, to require companies to conduct testing on chemical substances and mixtures; and (3) upon making certain findings, to take action to protect adequately against unreasonable risks such as by either prohibiting or limiting manufacture, processing, or distribution in commerce of chemical substances or by placing restrictions on chemical uses.[2] EPA maintains the TSCA Chemical Substance Inventory that currently lists over 84,000 chemicals that are or have been manufactured or processed in the United States; about 62,000 were already in commerce when EPA began reviewing chemicals in 1979.[3] Generally, TSCA's reporting requirements fall on the manufacturers (including importers), processors, and distributors of chemicals, rather than users of the chemicals.[4]

According to EPA, some of the chemicals on the TSCA Chemical Substance Inventory are used in oil and gas exploration and production. For example, in response to our request, EPA identified several

---

[1]Pub. L. No. 94-469, 90 Stat. 2003 (1976) (codified as amended at 15 U.S.C. §§ 2601 - 2692 (2012)). Hereinafter, references to TSCA are as amended. TSCA addresses those chemicals manufactured or imported into the United States, but it generally excludes certain substances, such as pesticides that are regulated under the Federal Insecticide, Fungicide, and Rodenticide Act, and any food, food additive, drug, cosmetic, or device regulated under the Federal Food, Drug, and Cosmetics Act.

[2]For example, prior to requiring testing under section 4, the act requires EPA to either make findings regarding the risk of injury to health or the environment or findings regarding human exposure, as well as findings regarding the sufficiency of existing data and that testing with respect to such effects is necessary to develop needed data. TSCA § 4(a), 15 U.S.C. § 2603(a) (2012).

[3]See http://www.epa.gov/oppt/existingchemicals/pubs/tscainventory/basic.html#background; GAO, *Chemical Regulation: Options for Enhancing the Effectiveness of the Toxic Substances Control Act*, GAO-09-428T (Feb. 26, 2009).

[4]See, e.g., TSCA § 8(a), (c), (d), 15 U.S.C. § 2607(a), (c), (d) (2012). Regulations also require users to take actions under the hazard communication provisions for certain substances in the workplace. See 40 C.F.R. §§ 721.3, 721.72 (2012).

chemicals on the FracFocus[5] list of "chemicals used most often" which are on the TSCA inventory.[6] These examples, which EPA chose as representative of different product function categories, are as follows:

- Hydrochloric acid – Acid;

- Peroxydisulfuric acid, ammonium salt – Breaker;

- Ethanaminium, 2-hydroxy-N,N,N-trimethyl-, chloride (1:1) – Clay Stabilizer;

- Methanol – Corrosion Inhibitor; and

- 2-Propenamide, homopolymer – Friction Reducer.

As part of EPA's *Study on the Potential Impacts of Hydraulic Fracturing on Drinking Water Resources*, EPA is currently analyzing information provided by nine hydraulic fracturing service companies, including a list of chemicals the companies identify as used in hydraulic fracturing operations. EPA officials said that they expect most of these chemicals disclosed by the service companies to appear on the TSCA inventory list, provided that chemicals are not classified solely as pesticides. EPA does not expect to be able to compare the list of chemicals provided by the nine hydraulic fracturing service companies to the TSCA inventory until the release of a draft report of the *Study on the Potential Impacts of Hydraulic Fracturing on Drinking Water Resources* for peer review, expected in late 2014. For those chemicals that are listed, some hydraulic fracturing service companies may be manufacturers, processors, or distributors, and could be subject to certain TSCA reporting provisions.

---

[5]FracFocus is the national hydraulic fracturing chemical registry managed by the Ground Water Protection Council and Interstate Oil and Gas Compact Commission. The Ground Water Protection Council is a national association of state groundwater and underground injection control agencies whose mission is to promote the protection and conservation of groundwater resources for all beneficial uses, recognizing groundwater as a critical component of the ecosystem. The Interstate Oil and Gas Compact Commission is a multi-state government agency that promotes the conservation and efficient recovery of domestic oil and natural gas resources while protecting health, safety, and the environment. http://www.fracfocus.org

[6]See http://www.epa.gov/oppt/existingchemicals/pubs/tscainventory, http://fracfocus.org/chemical-use/what-chemicals-are-used.

On August 4, 2011, Earthjustice and 114 others filed a petition with EPA asking the agency to exercise TSCA authorities and issue rules to require manufacturers, processors, and distributors of chemicals used in oil and gas exploration or production to develop and/or provide certain information.[7] The petition asserts that more than 10,000 gallons of such chemicals may be used to fracture a single well.[8] EPA denied the portion of the petition requesting that EPA issue a TSCA section 4 rule to require identification and toxicity testing of chemicals used in oil and gas exploration or production, stating that the petition did not set forth facts sufficient to support the findings required for such test rules.[9]

The petition also requested that EPA issue new rule(s) under TSCA section 8 to require, for these chemicals, maintenance and submission of various records, call-in of records of allegations of significant adverse reactions, and submission of all existing not previously reported health and safety studies.[10] EPA granted the section 8(a) and 8(d) portions of the petition in part, stating that the agency believes "there is value in initiating a proposed rulemaking process under TSCA authorities to obtain data on chemical substances and mixtures used in hydraulic fracturing," but denying them so far as they concern other chemical substances used in oil and gas exploration and production but not in hydraulic fracturing.[11]

EPA is drafting an Advance Notice of Proposed Rulemaking for the section 8(a) and (d) rules. As of August 31, 2012, EPA has not released a

---

[7]Earthjustice et al., Letter to Lisa P. Jackson, EPA Administrator, re: Citizen Petition under Toxic Substances Control Act Regarding the Chemical Substances and Mixtures Used in Oil and Gas Exploration or Production, Aug. 4, 2011. See also
http://www.epa.gov/oppt/chemtest/pubs/petitions.html#petition10

[8]Earthjustice et al., Letter to Lisa P. Jackson, EPA Administrator, re: Citizen Petition under Toxic Substances Control Act Regarding the Chemical Substances and Mixtures Used in Oil and Gas Exploration or Production, 2, Aug. 4, 2011.

[9]Stephen A. Owens, Assistant Administrator EPA, Letter to Deborah Goldberg, Earthjustice, Re: TSCA Section 21 Petition Concerning Chemical Substances and Mixtures Used in Oil and Gas Exploration or Production, Nov. 2, 2011.

[10]Earthjustice et al., Letter to Lisa P. Jackson, EPA Administrator, re: Citizen Petition under Toxic Substances Control Act Regarding the Chemical Substances and Mixtures Used in Oil and Gas Exploration or Production, Aug. 4, 2011.

[11]Stephen A. Owens, Assistant Administrator EPA, Letter to Deborah Goldberg, Earthjustice, Re: TSCA Section 21 Petition Concerning Chemical Substances and Mixtures Used in Oil and Gas Exploration or Production, Nov. 23, 2011.

publication date for this proposed rulemaking. EPA also intends to convene a stakeholder process to gather additional information for use in developing a proposed rule, and "to develop an overall approach that would minimize reporting burdens and costs, take advantage of existing information, and avoid duplication of efforts." EPA officials said that the agency will consider, among other things, how to address confidential business information as it develops the proposal. A TSCA section 8(a) rule, once issued, may require reporting, insofar as known or reasonably ascertainable, of such chemical information as chemical names, molecular structure, category of use, volume, byproducts, existing environmental and health effects data, disposal practices, and worker exposure.[12] Regulations promulgated under TSCA section 8(d) are to require submission to EPA of reasonably ascertainable health and safety studies.[13]

TSCA provides EPA with certain enforcement authorities. For example, EPA may impose a civil penalty for certain violations of TSCA,[14] such as failing to comply with requirements to notify and provide certain information to EPA before manufacturing a new chemical,[15] or by using for commercial purposes a chemical substance that the user had reason to know was manufactured, processed, or distributed in violation of such requirements, among other things.[16]

---

[12]TSCA § 8(a)(1)-(2), 15 U.S.C. § 2607(a)(1)-(2) (2012).

[13]TSCA § 8(d), 15 U.S.C. § 2607(d) (2012).

[14]TSCA §§ 16(a)(1), 15, 15 U.S.C. §§ 2615(a)(1), 2614 (2012).

[15]TSCA § 5(a)(1), 15 U.S.C. § 2604(a)(1) (2012). See also 40 C.F.R. pt. 720 (2012).

[16]TSCA § 15(2), 15 U.S.C. § 2614(2) (2012).

Appendix IX: Selected State Requirements

All six states we reviewed have state agencies responsible for implementing and enforcing environmental and public health requirements, which include overseeing oil and gas development (see table 11). In five of the six states we reviewed, this responsibility is split primarily between two different agencies. In general, one of these agencies has primary responsibility for regulating oil and gas development activities such as drilling that occur on the well pad and for managing and disposing of certain wastes generated on-site, while the other agency has a broader mandate for implementing and enforcing environmental or public health requirements, some aspects of which may affect oil and gas development. For example, the Colorado Oil and Gas Conservation Commission regulates activities such as drilling, hydraulic fracturing, and disposal of produced water in Class II UIC wells, while the Colorado Department of Public Health and Environment regulates discharges to surface waters, commercial solid waste facilities, and certain air emissions. In contrast, oil and gas development in Pennsylvania is primarily governed by one agency—the Pennsylvania Department of Environmental Protection.

Table 11: Primary State Agencies Responsible for Regulating Oil and Gas Development in Six States

State	State regulatory agencies
Colorado	Colorado Oil and Gas Conservation Commission
	Colorado Department of Public Health and Environment
North Dakota	North Dakota Industrial Commission, Oil and Gas Division
	North Dakota Department of Health
Ohio	Ohio Department of Natural Resources, Division of Oil and Gas Resource Management
	Ohio Environmental Protection Agency
Pennsylvania	Pennsylvania Department of Environmental Protection, Office of Oil and Gas Management
Texas	Texas Railroad Commission, Oil and Gas Division
	Texas Commission on Environmental Quality
Wyoming	Wyoming Oil and Gas Conservation Commission
	Wyoming Department of Environmental Quality

Source: GAO analysis of state information.

This appendix presents information about state statutory and regulatory requirements in the areas of siting and site preparation (see table 12); drilling, casing, and cementing (see table 13); hydraulic fracturing (see table 14); well plugging (see table 15); site reclamation (see table 16);

waste management in pits (see table 17); waste management through underground injection (see table 18); and managing air emissions (see table 19). Requirements presented in the following tables have been summarized mainly from state regulations, though references to state statutes are included in certain circumstances.[1,2,3]

Table 12: Selected State Requirements—Siting and Site Preparation

Identification or testing of water wells prior to drilling of production wells

CO   Testing requirements apply in certain circumstances. Specifically:

- *Coalbed methane wells.* If a conventional or plugged well exists within ¼ mile of a proposed coalbed methane well, the two closest water wells within a ½ mile must be sampled, if accessible. Wells must be tested for all major cations and anions, total dissolved solids, iron, manganese, selenium, nitrates, and nitrites, dissolved methane, field pH, sodium adsorption ration, presence of bacteria (iron related, sulfate reducing, slime, and coliform), specific conductance, and hydrogen sulfide. If there are no conventional or plugged wells within ¼ mile, or if access is denied to such wells, then a water well within ¼ mile, or, failing that, within ½ mile shall be selected. Post-completion sampling must be performed for the same substances within 1 year after completion of the well and repeated 3 and 6 years thereafter, or in accordance with field rules. The state may require further sampling at any time in response to complaints from well owners.

- *Wells in surface water supply areas.[a]* For new operations in surface water supply areas, pre-and post- drilling surface water samples must be taken for a number of substances, including pH, total dissolved solids, benzene, toluene, ethylbenzene, xylenes and metals, from streams immediately downgradient from the location. Different requirements apply to operations at locations that were in existence prior to the Spring of 2009 depending on whether new surface disturbance occurs at the site, and how much. *2 Colo. Code Regs. § 404-1(608, 317B) (2012).*

[1]References to state laws are included where no state regulation, or no detailed state regulation, exists; where the law was cross-referenced by a state regulation; or where interviews with state officials drew our attention to requirements in state law. These tables do not include state policy or practice; in some cases, states may address topic areas covered in these tables through processes that are not formally noted or comprehensively described in their regulations, such as the permitting process. The absence of a requirement in a particular state does not reflect any judgment on our part that a state should have such a requirement.

[2]In summarizing state rules, references to specific state officials and forms have been omitted except where those details are crucial for understanding the provision. In many cases, the requirements presented are default requirements that may be varied with state approval. Not all such options are noted in these tables. Unless otherwise noted, requirements are those that apply to new operations.

[3]State regulations on oil and gas development contain a variety of technical terms that are often not defined in the regulations themselves or may be differently defined across states. We have attempted to provide standard definitions for such terms for ease of reading; unless otherwise noted, however, such definitions are dictionary or industry glossary definitions rather than regulatory definitions.

ND	There is no testing requirement, but if a domestic, livestock, or irrigation water supply with 1 mile of an oil or gas well site is disrupted or diminished in quality or quantity by drilling operations and a certified water quality and quantity test has been performed within 1 year prior to drilling, a person owning an interest in the property supplied by that water is entitled to recover from the operator the costs of repairs, alterations, or construction necessary to deliver water of the original quality and quantity. Prima facie evidence of injury under this section may be established by a showing that the mineral developer's drilling operations penetrated or disrupted an aquifer in such a manner as to cause a diminution in water quality or quantity within the distance limits imposed by this section. *N.D. Cent. Code § 38-11.1-06 (2012).*
OH	The well owner must sample all water wells within 300 feet of proposed well locations in urbanized areas, and within 1,500 feet of proposed horizontal wells in any area, prior to drilling under the guidelines provided in the division's best management practices for predrilling water sampling manual.[b] The chief may require modification of this distance if determined necessary to protect water supplies or site conditions may warrant. *Ohio Rev. Code Ann. § 1509.06 (2012).*
PA	There is a rebuttable presumption that pollution occurring within 1000 feet and 6 months after completion of drilling or alteration of a conventional well and 2500 feet and 12 months after completion, drilling, stimulation or alteration, whichever is later, of an unconventional well was caused by the operator. Operators can defend against the presumption if they have predrilling tests showing that the problems predated drilling. The state does not specify substances for which wells must be tested. *25 Pa. Code § 78.51(2012); 58 Pa. Cons. Stat. § 3218 (2012).*
TX	No requirements identified in regulations or in statutes.
WY	An application for a permit to drill or deepen a well must identify all water supply wells permitted by the state within ¼ mile of the land unit within which the well is located, and the depth from which water is being appropriated. Owner/operators must also keep records on all formations penetrated and the content and quality of oil, gas, or water in each formation tested. *055-000-003 Code Wyo. R. §§ 8, 20 (2012).*

Required setbacks from water sources

CO	Special rules apply to new well sites depending on whether the site is located in one of three buffer zones surrounding surface water supply areas. Operations may not occur within the innermost buffer zone unless a variance is granted, the Department of Health and Environment is consulted, and appropriate conditions are placed on the operation. *2 Colo. Code Regs. § 404-1(317B) (2012).*
ND	Well sites and associated production facilities shall not be located in, or hazardously near, bodies of water or block natural drainages. *N.D. Admin. Code 43-02-03-19 (2012).*
OH	The location of a new well or a new tank battery of a well shall not be within 50 feet of a stream, river, watercourse, water well, pond, lake, or other body of water. However, the state may authorize a new well or tank battery to be located within 50 feet of such bodies of water if necessary to reduce impacts to the owner of the land or to protect public safety or the environment. *Ohio Rev. Code Ann. § 1509.021 (2012).*
PA	Conventional wells may not be drilled within 200 feet, and unconventional wells may not be drilled within 500 feet, of water wells without written owner consent. Unconventional wells may not be drilled within 1,000 feet of certain water supplies used by a water purveyor without written purveyor consent. If consent is not obtained, the operator may receive a variance if it cannot otherwise access its mineral rights and demonstrates that additional protective measures will be utilized. Conventional wells may not be drilled within 100 feet of certain other bodies of water, such as springs.[c] Unconventional wells may not be drilled within 300 feet of the same bodies of water and wetlands and the edge of the disturbed area associated with the well has to be at least 100 feet from the same water bodies and wetlands. *58 Pa. Cons. Stat. § 3215 (2012).*
TX	No requirements identified in regulations or in statutes.
WY	Generally, pits, wellheads, pumping units, tanks, and treaters shall be no closer than 350 feet from water supplies. *055-000-003 Code Wyo. R. § 22 (2012).*

	Erosion control, site preparation, surface disturbance minimization, and stormwater management
CO	• *Erosion control, site preparation, surface disturbance minimization.* Operators must separate excavated soil by horizon, store it separately and note locations to facilitate subsequent reclamation. On crop land, segregation must be to the shallower of 6 feet or bedrock. Elsewhere, operators must separate the topsoil horizon or the top 6 inches, whichever is deeper. If soil horizons are too rocky or thin to segregate, the topsoil shall be segregated and stored to the extent possible. Remaining soils on crop land shall be segregated down to the shallower of 3 feet or bedrock. Stockpiled soils shall be protected from contamination, compaction, and, as practicable, erosion. Best management practices to prevent weeds and maintain microbial activity shall be implemented. The drill pad shall minimize total disturbance consistent with safe operation and shall be on the most level location possible. If not avoidable, deep vertical cuts and steep long fill slopes shall be constructed to the least slope practical. Where feasible, directional drilling shall be used to reduce cumulative impacts and adverse impacts on wildlife. Well sites, production facilities, pipelines, and access roads shall be located, adequately sized, constructed, and maintained so as to reasonably control dust and minimize erosion, alteration of natural features, removal of surface materials, and degradation due to contamination. To the extent practicable, operators shall avoid or minimize impacts to wetlands and riparian habitats and shall consolidate facilities to minimize adverse impacts to wildlife resources, including fragmentation of wildlife habitat, as well as cumulative impacts. Existing roads shall be used to the greatest extent practicable to avoid erosion and minimize land disturbance. Roads shall be engineered to avoid or minimize impacts to riparian areas or wetlands. Unavoidable impacts shall be mitigated. Where feasible and practicable, road crossings of streams shall allow fish passage, operators are encouraged to share access roads in developing a field; roads shall be routed to complement other land usage, and vehicles shall not travel off-road. • *Stormwater management.* Operators must obtain a construction stormwater permit from the Department of Public Health and Environment and must develop a postconstruction stormwater program upon termination of the permit unless the site has a slope of less than 5% and has low erosion risk. All operators must implement and maintain site-specific best management practices to control stormwater runoff in a manner that minimizes erosion, transport of sediment off-site, and site degradation. Operators must select additional best management practices as part of their postconstruction stormwater program that address potential sources of pollution that may reasonably be expected to affect the quality of discharges associated with the ongoing operation of production facilities during the postconstruction and reclamation operation of the facilities. *2 Colo. Code Regs. § 404-1(1002) (2012).*
ND	• *Erosion control, site preparation, surface disturbance minimization.* In the construction of a drill site, access road, and all associated facilities, the topsoil shall be removed, stockpiled, and stabilized or otherwise reserved for use when the area is reclaimed. "Topsoil" means the first 8 inches of suitable plant growth material on the surface. Soil stabilization and materials to be used on-site, as well as access roads or associated facilities must have approval from the director before application. When necessary to prevent pollution of the land surface and freshwaters, the director may require the drill site to be sloped and diked. *N.D. Admin. Code 43-02-03-19 (2012).* • *Stormwater management.* Oil and gas construction activity that disturbs 5 or more acres is authorized under the terms of a general permit covering stormwater discharge that requires the development of a stormwater management program and implementation of best management practices.
OH	• *Erosion control, site preparation, surface disturbance minimization.* Site construction shall comply with the state's best management practices for oil and gas well site construction manual.[b] Site clearing and surface effects shall be minimized. During any phase of operation in urbanized areas, to minimize off-site sedimentation, erosion and to control the surface flow of water, the well owner or his representative must also follow the best management practices for oil and gas well site construction manual. Best management practices and design standards other than those provided by the state may be used if the alternative minimizes erosion to the same degree as the state procedure. *Ohio Admin. Code Ann. 1501:9-1-02, -07 (2012).*
PA	Operators proposing activities that will disturb 5,000 square feet or more or that have the potential to discharge to a high-quality or exceptional value water must develop and implement a written erosion and sedimentation plan. Operators proposing oil and gas activities that involve 5 acres or more of earth disturbance shall obtain an Erosion and Sedimentation permit prior to commencing the earth disturbance activity. During and after earth moving or soil disturbing activities, the operator must design, implement, and maintain best management practices relating to erosion and sediment control and postconstruction stormwater management. An operator may not commence drilling activities until the state has inspected the unconventional well site after the installation of erosion and sediment control measures. *25 Pa. Code §§ 78.53, 102.4, 102.5 (2012); 58 Pa.Cons.Stat. § 3258 (2012).*

TX	•	*Stormwater management.* Where required by federal law, discharges of stormwater associated with industrial and construction activities associated with the exploration, development, or production of oil or gas must be authorized by the EPA and the state, as applicable. Under federal law, EPA cannot require a permit for discharges of storm water from "field activities or operations associated with oil and gas exploration, production, processing, or treatment operations, or transmission facilities" unless the discharge is contaminated by contact with any overburden, raw material, intermediate product, finished product, byproduct, or waste product located on the site of the facility. Under state regulations, the Texas Railroad Commission prohibits operators from causing or allowing pollution of surface or subsurface water. Operators are encouraged to implement and maintain best management practices to minimize discharges of pollutants, including sediment, in stormwater to help ensure protection of surface water quality during storm events. *16 Tex. Admin. Code § 3.30 (2012).*
WY	•	*Erosion control, site preparation, surface disturbance minimization.* Where practical, topsoil must be stockpiled during construction for use in reclamation.
	•	*Stormwater management.* A permit is required for stormwater discharges from all construction activities disturbing 1 or more acres. These permits require the operator to develop a stormwater management program, including best management practices, which can be reviewed by the Wyoming Department of Environmental Quality. *055-000-003 Code Wyo. R. §§ 7, 17; 055-000-004 Code Wyo. R. § 1; 020-080-002 Code Wyo. R. § 6 (2012).*

Source: GAO analysis of state information.

[a]Surface water supply areas are certain streams that are suitable for or could become sources of drinking water that are within 5 miles upstream of a surface water intake.

[b]Best Management Practices for Pre-drilling Water Sampling Manual, available at http://www.ohiodnr.com/oil/watersampling_bmp/tabid/23361/Default.aspx The manual requires testing for barium, calcium, iron, magnesium, potassium, sodium, chloride, conductivity, pH, sulfate, alkalinity, and total dissolved solids.

[c]Specifically, no well site may be prepared or well drilled within 100 feet from any solid blue lined stream, spring or body of water as identified on the most current 7 and one-half minute topographic quadrangle map of the United States Geological Survey.

Table 13: Selected State Requirements—Drilling, Casing, and Cementing

Requirements relating to cementing/casing plans

CO	The casing[a] program adopted for each well must be so planned and maintained as to protect any potential oil or gas bearing horizons penetrated during drilling from infiltration of injurious waters from other sources, and to prevent the migration of oil, gas, or water from one horizon to another, which may result in the degradation of groundwater. *2 Colo. Code Regs. § 404-1(317) (2012).*
ND	The proposed casing program, including size and weight of casing, the depth at which each casing string[b] is to be set, the proposed pad layout including cut and fill diagrams, and the proposed amount of cement to be used, including the estimated top of cement, must be submitted with the application for permit to drill. *N.D. Admin. Code 43-02-03-16 (2012).*
OH[c]	A casing and cementing plan must be submitted as part of an application for permit to drill. The plan must show how the owner proposes to drill and construct the well with the best available geologic information in the vicinity of the proposed wellbore and with the requirements of state well construction rules. The plan must include, at least, the name and anticipated depth of all zones to be tested or produced; the estimated total depth of the wellbore; the anticipated diameter of each wellbore segment; the proposed casing type, outside diameter, and setting depth for each proposed casing string; proposed cement volumes for each casing string; and whether the owner plans to stimulate any permitted hydrocarbon zone by hydraulic fracturing. The casing and cementing plans in the approved permit are understood to be estimates based upon the best available geologic information prior to drilling. *Ohio Admin. Code Ann. 1501:9-1-02 (2012).*
PA	The operator must prepare a casing and cementing plan showing how the well will be drilled and completed. Upon request, the operator must provide a copy of the plan to the state for approval. The plan must include information such as the anticipated depth and thickness of any producing formation; expected pressures; anticipated fresh groundwater zones and the method or information by which the depth of the deepest fresh groundwater was determined; casing type; whether the casing is new or used; depth, diameter, wall thickness and burst pressure rating; cement type, yield, additives and estimated amount; estimated location of centralizers; proposed borehole conditioning procedures; and any alternative methods or

	materials required by DEP as a condition of well permit. *25 Pa. Code § 78.83a (2012).*
TX	Texas rules do not require the development or approval of a casing plan unless an operator proposes an alternative method of freshwater protection than those prescribed by rule. *16 Tex. Admin. Code § 3.13 (2012).*
WY	An application for permit to drill must include the proposed casing program including size, anticipated setting depths, American Petroleum Institute grade, weight per foot, burst pressure, tensile strength for both body and joint, yield pressure, if new or used casing is planned for the well, and any other information required by the state. *055-000-003 Code Wyo. R. § 8 (2012).*

Placement of surface casing relative to groundwater zones

CO	Where pressure and formations are unknown, surface casing[d] shall be run below all known or reasonably estimated utilizable domestic freshwater levels. Where subsurface conditions are known, surface casing shall be run to a depth sufficient to protect all freshwater. Where freshwater aquifers are so deep that it is impractical/uneconomical to set casing to the required depth, intermediate and/or production casing may be stage cemented to isolate the aquifers. *2 Colo. Code Regs. § 404-1(317) (2012).*
ND	Casing must be properly cemented at sufficient depths to adequately protect and isolate all formations containing water, oil, or gas or any combination of these. The surface casing shall be set and cemented at a point not less than 50 feet below the base of the Fox Hills formation. *N.D. Admin. Code 43-02-03-21 (2012).*
OH	An owner shall set and cement surface casing at least 50 feet below the base of the deepest underground source of drinking water (USDW), or at least 50 feet into competent bedrock, whichever is deeper and as specified by the permit, unless otherwise approved by the state. In areas where bedrock USDWs cannot be mapped and where groundwater resources can be developed in valley-fill aquifers, surface casing shall be cemented at least 100 feet below the base of the valley-fill aquifer for any well within 1,000 feet of the 100 year floodplain. In other areas where bedrock USDWs cannot be mapped, surface casing shall be set and cemented to at least 300 feet deep or at least 100 feet below the deepest local perennial stream base. As an alternative where bedrock USDWs cannot be mapped, surface casing shall be set to at least 50 feet below the base of the lowest spring or deepest water well within 500 feet or if there are no such springs or water wells, conductor casing shall be set and cemented to at least 100 feet. After conductor casing is set through the deepest useable water zone and cemented to surface, the owner must set and cement to surface a surface casing string through water zones that may include brackish or brine bearing zones. This casing string shall be set and cemented to surface before the owner drills into potential flow zones that can reasonably be expected to contain hydrocarbons in commercial quantities. Other alternative methods of protecting USDWs may be approved upon written application to the state. *Ohio Admin. Code Ann. 1501:9-1-08 (2012).*
PA	Surface casing must be set to approximately 50 feet below deepest fresh groundwater or at least 50 feet into consolidated rock, whichever is deeper. Generally, surface casing may not be set more than 200 feet below the deepest fresh groundwater unless necessary to set casing in consolidated rock. Cement placed behind the surface casing must be set to a minimum designed strength of 350 pounds per square inch (psi) at the casing seat. The cement placed at the bottom 300 feet of the surface casing must constitute a zone of critical cement and achieve a 72-hour compressive strength of 1,200 psi, and the free water separation may be no more than 6 milliliters of cement. If the surface casing is less than 300 feet, the entire cemented string constitutes a zone of critical cement. *25 Pa. Code §§ 78.83, .85 (2012).*
TX	Surface casing must be set to protect all usable-quality water strata, as defined by the Texas Commission on Environmental Quality (TCEQ). Before drilling any well where no field rules are in effect or in which surface casing requirements are not specified in the applicable field rules, an operator shall obtain a letter from the TCEQ stating the protection depth. In no case, however, is surface casing to be set deeper than 200 feet below the specified depth without prior approval from the Texas Railroad Commission. *16 Tex. Admin. Code § 3.13 (2012).*
WY	Surface casing shall be run below all known or reasonably estimated utilizable groundwater. Generally, surface casing shall be set at a minimum of 100 to 120 feet below the depth of any state permitted wells designated for domestic, stock water, irrigation, or municipal use within a minimum of 1/4 mile. A coalbed methane well with a groundwater appropriation permit is exempt from this requirement. *055-000-003 Code Wyo. R. § 22 (2012).*

Prescribed cementation techniques for surface casing

CO	Pump and plug, displacement, or other approved method. *2 Colo. Code Regs. § 404-1(317) (2012).*
ND	Pump and plug method or other state-approved method. *N.D. Admin. Code 43-02-03-21 (2012).*
OH	Sufficient cement shall be used to fill the annular space outside the casing from the seat to the ground surface or to the bottom of the cellar. If cement is not circulated to the ground surface or the bottom of the cellar, and the top of cement cannot be measured from surface, the owner shall notify the state and perform certain tests to determine the nature of the deficiency and shall obtain approval for additional cementing operations. *Ohio Admin. Code Ann. 1501:9-1-08 (2012).*
PA	Operator shall permanently cement surface casing by placing cement in casing and displacing it into annular space between wall of hole and outside of casing. *25 Pa. Code § 78.83 (2012).*
TX	Surface casing shall be cemented by the pump and plug method. The producing string of casing shall be cemented by the pump and plug method, or other commission-approved method. Alternative surface casing programs may be approved upon written application. *16 Tex. Admin. Code § 3.13 (2012).*
WY	Pump and plug, displacement, or other approved method. *055-000-003 Code Wyo. R. § 22 (2012).*

Requirement for cement waiting period and/or integrity tests

CO	Surface and intermediate casing cement must achieve a minimum compressive strength of 300 psi after 24 hours and 800 psi after 72 hours measured at 95°F and 800 psi. Cement placed behind surface and intermediate casing shall be allowed to set a minimum of 8 hours, or until 300 psi calculated compressive strength is developed, whichever occurs first, prior to commencing drilling operations. Cement placed behind production casing shall achieve a minimum compressive strength of at least 300 psi after 24 hours and 800 psi after 72 hours measured at 95°F and 800 psi. Cement placed behind production casing shall be allowed to set 72 hours, or until 800 psi calculated compressive strength is developed, whichever occurs first, prior to any completion operation. Installed production casing shall be adequately pressure-tested for the conditions anticipated to be encountered during completion and production. *2 Colo. Code Regs. § 404-1(317) (2012).*
ND	All strings of surface casing shall stand cemented under pressure for at least 12 hours before drilling the plug or initiating tests. Surface casing strings must be allowed to stand under pressure until the tail cement has reached a compressive strength of at least 500 psi. All filler cements must reach a compressive strength of at least 250 psi within 24 hours and at least 350 psi within 72 hours. Compressive strength on surface casing cement shall be calculated at 80° F. Production or intermediate casing strings must be allowed to stand under pressure until the tail cement has reached a compressive strength of at least 500 psi. All filler cements utilized must reach a compressive strength of at least 250 psi within 24 hours and at least 500 psi within 72 hours, although in any horizontal well performing a single stage cement job from a measured depth of greater than 13,000 feet, the filler cement utilized must reach a compressive strength of at least 250 psi within 48 hours and at least 500 psi within 96 hours. After cementing, each casing string shall be tested by application of pump pressure of at least 1,500 psi. If, at the end of 30 minutes, this pressure has dropped 150 psi or more, the casing shall be repaired and tested in the same manner again. Further work shall not proceed until a satisfactory test has been obtained. The casing in a horizontal well may be tested by use of a mechanical tool set near the casing shoe[e] after the horizontal section has been drilled. *N.D. Admin. Code 43-02-03-21 (2012).*
OH	Cemented conductor, mine, and surface casing strings shall remain static until all cement has reached a compressive strength of at least 500 psi before drilling the plug, or initiating a test. The tail cement for all intermediate and production casing and liners shall remain static until the cement has reached a compressive strength of at least 500 psi before drilling out the plug or initiating a test. Tail cement shall have a 72-hour compressive strength of at least 1,200 psi. Lead cements with volume extenders may be used to seal these strings but in no case shall the cement have a compressive strength of less than 100 psi at the time of drill out nor less than 250 psi 24 hours after being placed. Cement mixtures for which published performance data are not available shall be tested by the owner or service company and approved prior to usage. Tests shall be made on representative samples of the basic mixture of cement and additives used, using distilled water or potable tap water for preparing the slurry. The tests shall be conducted using the equipment and procedures established in the American Petroleum Institute publication "RP 10 B-2 Recommended Practice for Testing Well Cements." Test data showing competency of a proposed cement mixture to meet the above requirements shall be furnished to the inspector prior to the cementing operation. To determine that the minimum compressive strength has been obtained, the owner shall use the typical performance data for the particular cement mixture used in the well at the following temperatures and at atmospheric pressure: for conductor, mine string, and surface casing cement, the test temperature shall be 60°F; for intermediate and production casing cement, the test temperature shall be within 10°F of the formation equilibrium temperature of the cemented interval. *Ohio Admin. Code Ann. 1501:9-1-08 (2012).*

PA	After the casing cement is placed behind surface casing, the operator shall permit the cement to set to a minimum designed compressive strength of 350 psi at the casing seat. The cement placed at the bottom 300 feet of the surface casing must constitute a zone of critical cement and achieve a 72-hour compressive strength of 1,200 psi ,and the free water separation may be no more than 6 milliliters per 250 milliliters of cement. After any casing cement is placed and cementing is complete, casing generally may not be disturbed for at least 8 hours by certain activities, such as running drill pipe or other mechanical devices into or out of the wellbore with the exception of certain equipment used to determine the top of the cement. *25 Pa. Code 78.85 (2012).*
TX	When cementing any string of casing more than 200 feet long, before drilling the cement plug, the operator shall test the casing at a pump pressure in psi calculated by multiplying the length of the casing string by 0.2. The maximum test pressure required, however, unless otherwise ordered by the commission, need not exceed 1,500 psi. If, at the end of 30 minutes, the pressure shows a drop of 10% or more from the original test pressure, the casing shall be condemned until the leak is corrected. A pressure test demonstrating less than a 10% pressure drop after 30 minutes is proof that the condition has been corrected. Surface casing strings must be allowed to stand under pressure until the cement has reached a compressive strength of at least 500 psi in the zone of critical cement before drilling plug or initiating a test. The cement mixture in the zone of critical cement shall have a 72-hour compressive strength of at least 1,200 psi. Cement mixtures for which published performance data are not available must be tested by the operator or service company in accordance with the current API RP 10B.^{f} Test data showing competency of a proposed cement mixture must be furnished to the commission prior to cementing. To determine that the minimum compressive strength has been obtained, operators shall use typical performance data for the cement in the well. *16 Tex. Admin. Code § 3.13 (2012).*
WY	Unless otherwise provided by specific order of the Oil and Gas Conservation Commission for a particular well or wells or for a particular pool or parts thereof, cemented casing string shall stand under pressure until the cement at the shoe has reached a compressive strength of 500 psi. In addition, the American Petroleum Institute free-water separation for all cement slurries used shall average no more than 4 mL per 250 mL of cement. All cements used shall achieve a minimum compressive strength of 100 psi in 24 hours measured at 80° F. Testing for these properties shall be in accordance with accepted industry standards. *055-000-003 Code Wyo. R. § 22 (2012).*

Other measurement, record keeping, notification and/or inspection during cementing/casing process

CO	A cement bond log shall be run on all production casing or, in the case of a production liner, the intermediate casing when these strings are run. Open hole logs shall be run at depths that adequately verify the setting depth of surface casing and any aquifer coverage. *2 Colo. Code Regs. § 404-1(317) (2012).*
ND	If the annular space behind casing is not adequately filled with cement, the state must be notified immediately. Any well that appears to have defective casing or cementing, the operator shall report the defect to the director. Prior to attempting remedial work on any casing, the operator must obtain approval from the director and proceed with diligence to conduct tests, as approved or required by the director, to properly evaluate the condition of the wellbore and correct the defect. *N.D. Admin. Code 43-02-03-21, -22 (2012).*
OH	Generally, the state must receive 48 hours prior notice before the placement of surface casing. A 24 hours or less notification may be approved if prior communications have been initiated with state officials. The state must receive 24 hours notice prior to setting any casing or liner string and before commencing any casing cementing. Within 60 days after drilling to total depth, the owner shall file cement job logs with the state furnishing complete data documenting the cementing of all cemented casing strings. Each job log shall include the date cemented; the name of the cementing contractor; mix water temperature and pH; whether or not the wellbore circulated prior to cementing; certain measurements for the hole, casing, and other equipment; cement types, additives by percent of unit volume, volume of cement in stacks, cement yield per sack, average slurry density, slurry volume, and displacement volume; pumping rates, displacement pressure, and final circulating pressure prior to landing the plug; the time the latch-down or wiper plug landed; casing test pressure and final test pressure; whether or not cement circulated to surface; and volume of cement slurry circulated to the surface. *Ohio Admin. Code Ann. 1501:9-1-02, 1501:9-1-08 (2012).*
PA	Operator must notify state if cement used to cement surface casing is not circulated to the surface despite use of at least 120% of expected volume. Casing must undergo pressure testing, and the operator must notify the state at least 24 hours before conducting pressure testing on casing. The operator shall notify the state at least 1 day before cementing of the surface casing begins, unless the cementing operations begins within 72 hours of commencement of drilling. Unconventional well operators shall provide the state 24 hours notice prior to cementing all casing strings, conducting pressure tests of the production casing, stimulation, and abandoning or plugging an unconventional well. *25 Pa. Code §§ 78.83b, .84, .85 (2012); 58 Pa.Cons.Stat. § 3211 (2012).*

TX	Upon completion of the well, a cementing report must be filed with the commission furnishing complete data concerning the cementing of surface casing in the well. *16 Tex. Admin. Code § 3.13 (2012).*
WY	The state may require a well owner or operator to provide bond logs if there is a demonstrated reason to believe an inadequate cement job was performed. A cement bond log or cement evaluation tool must be run to verify adequate cement around surface casing and to evaluate cement integrity in each cemented zone for each cemented casing annulus in wells within a geologic area known as the Special Sodium Drilling Area. *055-000-003 Code Wyo. R. § 22 (2012).*

Requirements related to horizontal drilling

CO	If an operator intends to drill a horizontal or deviated wellbore utilizing controlled directional drilling methods, other than whipstocking[9] due to hole conditions, the plans shall accompany an application for permit to drill. The plat shall show the surface and bottom hole location. If the surface location is in a different section than the bottom hole location, a plat depicting each section is required. Additionally, the proposed directional survey including two wellbore deviation plots, one depicting the plan view and one depicting the side view, shall accompany the application. Within 30 days of completion, the operator shall submit a drilling completion report, with a copy of the directional survey coordinate listing and the wellbore deviation plots. The survey data shall be provided in a single analysis report with sufficient detail to determine the location of the wellbore from the base of the surface casing to the kick off point and from that point to total depth. Special spacing rules apply to horizontal wells in the Greater Wattenberg Area. *2 Colo. Code Regs.§§ 404-1(321, 318A) (2012).*
ND	A permit is required prior to drilling horizontally in an existing pool. A directional survey shall be made and filed with the director on any well utilizing a whipstock or any method of deviating the wellbore. Special permits may be obtained to drill directionally in a predetermined direction. If a request for a permit to directionally drill is denied, the director shall immediately tell the applicant why. The decision of the director may be appealed to the commission. Filler cement used to set production or intermediate casing in any horizontal well performing a single stage cement job from a measured depth of greater than 13,000 feet must reach a compressive strength of at least 250 psi within 48 hours and at least 500 psi within 96 hours. Casing in a horizontal well may be tested by use of a mechanical tool set near the casing shoe after the horizontal section has been drilled. *N.D. Admin. Code 43-02-03-16, -21, -25 (2012).*
OH	The maximum point at which a well penetrates the producing formation shall not vary unreasonably from the vertical drawn from the center of the hole at the surface, with the exception of approved directional drilling. Such approval must be in writing from the chief. For wells drilled horizontally, in the Marcellus shale, or deeper, intermediate casing shall be set through the Mississippian Berea sandstone or to 1,000 feet, whichever is greater, or as determined by the state. Production casing shall be cemented with sufficient cement to fill the annular space to a point at least 500 true vertical feet above the seat in an open-hole vertical completion or the uppermost perforation in a cemented vertical completion, or 1,000 feet above the kickoff point of a horizontal well. Liners may only be set and cemented as production casing in horizontal shale gas wells if approved by the chief. *Ohio Admin. Code Ann. 1501:9-1-02, -08 (2012).*
PA	No requirements identified in regulations or in statutes.[h]
TX	A permit for directionally deviating a well may be granted for a variety of reasons, including where it can be shown to be advantageous from the standpoint of mechanical operation to drill more than one well from the same surface location to reach the productive horizon at essentially the same positions as would be reached if the several wells were drilled from locations prescribed by the well spacing rules. Applications for directional deviation must specify surface and projected bottom hole locations. *16 Tex. Admin. Code § 3.11 (2012).*[i]
WY	For directional wells, an application for permit to drill must include a diagram showing the proposed direction of the deviation and the proposed horizontal distance between hole bottom and surface location. For horizontal wells, an application for permit to drill must include a diagram showing the wellbore path from the surface through the terminus of the lateral. The surface location and the proposed footage locations of both the initial penetration into the productive formation and the terminus of the lateral shall be recorded. For an application to drill a horizontal well, notice shall be given to owners within 1/2 mile of any point on the length of the wellbore. In the absence of any special state order, notice is not required for horizontal wells in federally supervised units or in American Petroleum Institute units provided that no portion of the horizontal interval is closer than 660 feet from a drilling or spacing unit boundary or uncommitted tract. Before beginning controlled directional drilling, other than whipstocking because of hole conditions, notice shall be filed and approval obtained. Such notice shall state: the depth; exact surface location of the wellbore; proposed direction of deviation; and proposed horizontal distance between the bottom of the hole and surface location. A directional survey must be filed within 30 days of completion. *055-000-003 Code Wyo. R. §§ 8, 25 (2012).*

Blowout preventer requirements

CO	Operators must take all necessary precautions for keeping a well under control while being drilled or deepened. Blowout preventer equipment is not generally required but must be used in high density areas or other areas specified by the state. In addition, the state may designate specific areas, fields or formations as requiring certain blowout prevention equipment. If used, blowout preventer equipment must be identified in the operator's application for permit to drill. The working pressure of any blowout preventer equipment shall exceed the anticipated surface pressure to which it may be subjected. *2 Colo. Code Regs. § 404-1(317, 603) (2012).*
ND	In all drilling operations, proper and necessary precautions shall be taken for keeping the well under control, including the use of a blowout preventer and high pressure fittings attached to properly cemented casing strings adequate to withstand anticipated pressures. *N.D. Admin. Code 43-02-03-23 (2012).*
OH	Blowout preventer equipment is not generally required, but casing integrity may be verified in conjunction with blowout preventer testing without a test plug. *Ohio Admin. Code Ann. 1501:9-1-08 (2012).*
PA	Blowout preventers are required (1) when drilling into an unconventional formation,[j] (2) when drilling out solid core hydraulic fracturing plugs to complete a well, (3) when anticipated well head pressures or natural open flows may result in a loss of well control, (4) where there is no prior knowledge of the pressures or natural open flows, (5) on wells drilled to at least 3,800 feet and penetrating the Onondaga horizon, and (6) when drilling within 200 feet of a building. The operator shall use pipe fittings, valves, and unions placed on or connected to the blowout prevention systems that have a working pressure capability that exceeds the anticipated pressures. All lines, valves, and fittings between the closing unit and the blowout preventer stack must be flame resistant and have a rated working pressure that meets or exceeds the requirements of the blowout preventer system. *25 Pa. Code § 78.72 (2012).*
TX	Wellhead assemblies shall be used on wells to maintain surface control of the well. Each component of the wellhead shall have a pressure rating equal to or greater than the anticipated pressure to which that particular component might be exposed during the course of drilling, testing, or producing the well. A blowout preventer or control head and other connections to keep the well under control at all times shall be installed as soon as surface casing is set. The equipment shall be of such construction and capable of such operation as to satisfy any reasonable test that may be required by the commission or its duly accredited agent. *16 Tex. Admin. Code § 3.13 (2012).*
WY	Blowout preventers and related equipment shall be installed and maintained during drilling in accordance with state rules unless changed upon hearing before the Oil and Gas Conservation Commission. Among other things, the requirements include that the working pressure rating of all blowout preventers and related equipment shall be based on known or anticipated subsurface pressure, geologic conditions, or accepted engineering practices, and shall equal or exceed the maximum anticipated pressure to be contained at the surface. In the absence of better data, the maximum anticipated surface pressure shall be determined by using a normal pressure gradient of 0.22 psi per foot and assuming a partially evacuated hole. *055-000-003 Code Wyo. R. § 23 (2012).*

Source: GAO analysis of state information.

[a]Casing is metal pipe used to line a well.

[b]A string is a column made up of connected pipe.

[c]In Ohio, the state may establish alternative well construction standards that are well-specific, field-specific, or play-specific by permit condition, to ensure protection of public health or safety or the environment. *Ohio Admin. Code Ann. 1501:9-1-08(B) (2012).*

[d]Surface casing is cemented into bedrock and serves to shut out shallow water formations and as a foundation for well control during drilling.

[e]A casing shoe is a cylinder or ring of hard steel with a cutting edge attached to the bottom of a string of well casing.

[f]American Petroleum Institute, "API RP 10B-2 (R2010) Recommended Practice for Testing Well Cements." July 2005.

[g]Whipstocking is a long wedge dropped into or placed in a well to deflect the drill to one side of some obstruction.

[h]Pennsylvania requires operators to provide information about the vertical and horizontal paths of unconventional wells through its permitting process.

[i]See also 16 Tex. Admin. Code § 3.86, relating to horizontal drainhole wells.

ʲPennsylvania law defines an unconventional formation as a geological shale formation existing below the base of the Elk Sandstone or its geologic equivalent stratigraphic interval where natural gas generally cannot be produced at economic flow rates or in economic volumes except by vertical or horizontal well bores stimulated by hydraulic fracture treatments or by using multilateral well bores or other techniques to expose more of the formation to the well bore. 58 Pa.Cons.Stat. § 2301 (2012).

Table 14: Selected State Requirements—Hydraulic Fracturing

Prior authorization/notice/inspection requirements

CO	Operators shall give at least 48 hours advance written notice to the Oil and Gas Conservation Commission of a hydraulic fracturing treatment at any well. The state must provide prompt electronic notice of such intention to the relevant local governmental designee. *2 Colo. Code Regs. § 404-1(316C) (2012).*
ND	No requirements identified in regulations or in statutes.
OH	The casing and cementing plan that must be submitted with an application for permit to drill must indicate whether the owner plans to stimulate any permitted hydrocarbon zone by hydraulic fracturing. The owner or the owner's representative must notify the state at least 24 hours before commencing the stimulation of a well. *Ohio Admin. Code Ann. 1501:9-1-02(A) (2012). Ohio Rev. Code Ann. § 1509.19 (2012).*
PA	Operators must give 24 hours notice prior to well stimulation. *58 Pa. C.S. § 3211 (2012).*
TX	No requirements identified in regulations or in statutes.
WY	An application for permit to drill must include a description of the anticipated stimulation program. An approved Application for Permit to Drill or Sundry Notice is required prior to the initiation of any well stimulation activity. *055-000-003 Code Wyo. R. §§ 8, 45 (2012).*

Requirements to disclose information on fracturing fluids

CO	Certain disclosures are required from vendors of hydraulic fracturing additives, companies that provide hydraulic fracturing services, and operators on whose wells hydraulic fracturing treatments are completed. Vendors and service providers must, with the exception of information claimed to be a trade secret, furnish to operators information on the total volume of water or the type and total volume of any other base fluid used in the hydraulic fracturing treatment; each hydraulic fracturing additive used in the hydraulic fracturing fluid along with its trade name, vendor, and a brief descriptor of its intended use; each chemical intentionally added to the base fluid, its maximum concentration, and its Chemical Abstracts Service (CAS) number, if applicable. Vendors and service providers must also provide any other information needed for operators to comply with operators' own disclosure requirements. Operators must post information to FracFocus, including: information on the makeup of hydraulic fracturing treatments as described above; operator name; date of hydraulic fracturing treatment; location and other identifying information for the well; and the true vertical depth of the well. If the specific identity of a chemical, its concentration, or both is/are claimed as a trade secret, the operator must so indicate on its submission to FracFocus and, as applicable, the vendor, service provider, or operator shall submit to the state a claim of entitlement to have the information withheld as a trade secret. Even if a chemical is claimed to be a trade secret, the operator must include in its submission the chemical family or other similar descriptor associated with such chemical. Information claimed to be a trade secret shall be provided to any health professional who provides a written statement of need for the information and executes a confidentiality agreement. Where a health professional determines that a medical emergency exists and the protected information is necessary for emergency treatment, the information shall be immediately disclosed upon a verbal acknowledgement that such information shall not be used for purposes other than the asserted health needs and shall otherwise be maintained as confidential. A written statement of need and a confidentiality agreement may be requested as soon as circumstances permit. Information so disclosed does not become publicly available. Information claimed to be a trade secret shall be provided to the Oil and Gas Conservation Commission upon written request from the Director of the Commission stating that such information is necessary to respond to a spill or release or a complaint from a person who may have been affected or aggrieved by a spill or release. The Director or designee may disclose the information to additional commission staff if such disclosure is necessary to allow staff to respond to the spill, release, or complaint, provided that such individuals shall not disseminate the information further. In addition, the Director may disclose such information to any commissioner, the relevant county public health director or emergency manager, or to the Colorado Department of Public Health and Environment's director of environmental programs upon request by that individual. The Colorado Department of Public Health and Environment's director of environmental programs, or his or her designee, may disclose such information

	to Colorado Department of Public Health and Environment staff members under the same terms and conditions as apply to the director. Information so disclosed does not become publicly available. *2 Colo. Code Regs. § 404-1(100, 205A) (2012)*.
ND	After performance of a hydraulic fracture stimulation, the owner, operator, or service company must post on FracFocus all elements made viewable by the FracFocus website.[a] *N.D. Admin. Code 43-02-03-27.1 (2012)*.
OH	Within 60 days after the completion of drilling or after a determination that a well is a dry or lost hole, a well completion record must be filed which designates: the trade name and the total amount of all products, fluids, and substances, and the supplier of each product, fluid, or substance—not including cement and its constituents and lost circulation materials—intentionally added to facilitate the drilling of any portion of the well until the surface casing is set and properly sealed or to stimulate the well. The owner shall identify each additive used and provide a brief description of the purpose for which the additive is used. For stimulated wells, the owner shall also include the maximum concentration of the additive used. In addition, the owner shall include a list of all chemicals, not including any information that is designated as a trade secret, intentionally added to all products, fluids, or substances and include each chemical's corresponding CAS number and the maximum concentration of each chemical. The owner shall obtain the chemical information, not including trade secrets, from the company that drilled or stimulated the well, provided drilling services at the well, or supplied the chemical; and the type and volume of any fluid, not including cement and its constituents or trade secret information, used to stimulate the well, the reservoir breakdown pressure, the method used for the containment of fluids recovered from the fracturing of the well, the methods used for the containment of fluids when pulled from the wellbore from swabbing the well, the average pumping rate of the well, and the name of the person that performed the well stimulation. In addition, the owner shall include a copy of the log from the stimulation of the well, a copy of the invoice for each of certain procedures and methods that were used on a well, and a copy of the pumping pressure and rate graphs. After a well is initially completed and stimulated and until the well is plugged, the owner shall report all materials placed into the formation to refracture, restimulate, or newly complete the well, in addition to the information required above, within 60 days. If there is a material listed in the disclosures for which the state does not have a material safety data sheet, the owner shall provide a copy to the state. Information must be submitted to the state on state prescribed forms, through FracFocus, or by other state-approved means. The state must post information obtained online. If a medical professional, in order to assist in the diagnosis or treatment of an individual who was affected by an incident associated with the production operations of a well, requests the exact chemical composition of each product, fluid, or substance and of each chemical component in a product, fluid, or substance that is designated as a trade secret pursuant, the person claiming the trade secret protection shall provide the information requested. A medical professional who receives such information shall keep it confidential and shall not disclose it for any purpose that is not related to the diagnosis or treatment. This requirement does not preclude a medical professional from making any report required by law or professional ethical standards. Companies may withhold the identity, amount, concentration, or purpose of a product, fluid, or substance or of a chemical component in a product, fluid, or substance as a trade secret. The state may not disclose any trade secret information. Anyone with an interest that is or may be adversely affected by a product, fluid, or substance or by a chemical component in a product, fluid, or substance may challenge a claim of trade secret protection. A well owner shall maintain records of chemicals placed in a well for at least 2 years after placement. The chief of the Oil and Gas Division in the Ohio Department of Natural Resources may inspect the records for non trade secret information at any time concerning any such chemical. For trade secret information, the well owner must disclose records to the state upon request if the information is necessary to respond to a spill, release, or investigation. However, the state shall not disclose the information that is designated as a trade secret. An owner is not required to report chemicals that occur incidentally or in trace amounts. *Ohio Rev. Code Ann. § 1509.10 (2012)*.
PA	Certain disclosures are required from vendors of hydraulic fracturing additives, companies that provide hydraulic fracturing services, and operators on whose wells hydraulic fracturing treatments are completed.[b] If the vendor, service provider or operator claims that the specific identity of a chemical or the concentration of a chemical, or both, are a trade secret or confidential proprietary information, the operator of the well must indicate that on the form submitted to FracFocus and submit a signed statement that the record contains a trade secret or confidential proprietary information. If a chemical is a trade secret, the operator shall include in its submission to FracFocus the chemical family or similar description associated with the chemical. Unless the information is entitled to protection as a trade secret or confidential proprietary information, information posted to FracFocus is public. A vendor, service company, or operator shall identify the specific identity and amount of any chemicals claimed to be a trade secret or confidential proprietary information to any health professional who executes a confidentiality agreement and provides a written statement of need for the information indicating that the information is needed for diagnosis or treatment of an individual who may have been exposed to a hazardous chemical, and that knowledge of the information will assist in the diagnosis or treatment. If a health professional determines that a medical emergency exists and the information claimed to be a trade secret or confidential proprietary information is necessary for emergency treatment, the vendor, service provider or operator shall immediately disclose the information upon a verbal

acknowledgment that the information may not be used for purposes other than the health needs asserted and that it will be maintained as confidential. The vendor, service provider, or operator may request, and the health professional shall provide, a written statement of need and a confidentiality agreement as soon as circumstances permit. Trade secret or confidential proprietary information must also be provided to the department, a public health official, an emergency manager, or a responder to a spill, release or a complaint if that information is needed for response. The department must prevent further disclosure of trade secrets or confidential proprietary information in accordance with other applicable state law. *58 Pa.Cons.Stat. §§ 3203, 3222.1 (2012).*

TX	Certain disclosures are required from suppliers of hydraulic fracturing additives, companies that provide hydraulic fracturing services, and operators on whose wells hydraulic fracturing treatments are completed. Suppliers and service providers must provide to operators information on each chemical ingredient intentionally added to the hydraulic fracturing fluid, including each additive used, its trade name, supplier, and a brief description of its intended use; each chemical ingredient for which a material safety data sheet would be required under federal regulation;[c] the actual or maximum concentration of each chemical ingredient listed under the last two clauses in percent by mass; all other chemical ingredients that were intentionally included in and used for creating hydraulic fracturing treatment(s) for the well; and the Chemical Abstracts Service (CAS) number for each chemical ingredient, if applicable. Operators must post information to FracFocus, including: information on the makeup of hydraulic fracturing treatments as described above; operator name; date of hydraulic fracturing treatment; location and other identifying information for the well; the true vertical depth of the well; and the total volume of water used in the hydraulic fracturing treatment(s) of the well or the type and total volume of the base fluid used in the hydraulic fracturing treatment(s), if something other than water. If the supplier or service company claims that the specific identity and/or CAS number or amount of any additive or chemical ingredient used in a treatment is entitled to protection as a trade secret under Texas law, it must provide the operator a written statement to that effect and must provide its contact information and the chemical family, unless that is also claimed as a trade secret, in which case only the properties and effects of the ingredient(s) must be disclosed. A claim that any of the same information is a trade secret must be included in the submission to FracFocus, along with the chemical family or other similar description and the contact information of the business claiming the trade secret. Information not protected as a trade secret is public information. A supplier, service company, or operator may not withhold information, including trade secrets, from any health professional or emergency responder who needs the information for diagnostic, treatment or other emergency response purposes. A supplier, service company, or operator must provide directly to a health professional or emergency responder all information in the person's possession that is required whether or not the information may be a trade secret. The person disclosing information must include, as soon as circumstances permit, a statement of the health professional's confidentiality obligation. In an emergency, the supplier, service company, or operator must provide the information immediately.[d] A health professional or emergency responder to whom information is disclosed must hold the information confidential, except that he or she may, for diagnostic or treatment purposes, disclose information to another health professional, emergency responder, or accredited laboratory. Such a person or entity must hold the information confidential, and the disclosing person must include with the disclosure, or in a medical emergency, as soon as circumstances permit, a statement of the recipient's confidentiality obligation. Certain parties may challenge a claim of trade secret protection, including the landowner on whose property the relevant wellhead is located; the landowner who owns real property adjacent to property described above; and a department/agency with jurisdiction over a matter to which the claim is relevant. *16 Tex. Admin. Code § 3.29 (2012).*
WY	• The Application for Permit to Drill must include a description of any anticipated stimulation program, including the base stimulation fluid and its source, the chemical additives and proposed concentrations to be mixed, identified by additive type. Specifically, owners or operators must provide information on the stimulation fluid identified by additive type, the chemical compound name and CAS number, and the proposed rate or concentration for each additive. Upon prior request on certain forms and/or by written letter to the state justifying and documenting the nature and extent of the proprietary information, confidentiality protection shall be provided consistent with the Wyoming Public Records Act for trade secrets, privileged information and confidential commercial, financial, geological, or geophysical data furnished by or obtained from any person. Reports must generally be submitted to the state within 30 days of completion of certain activities, including formation fracturing, which present a detailed account of the work done and the manner in which it was performed, including the quantity of sand, crude, chemical, or other materials employed in the operation. • Following well stimulation, the owner, operator, or service company must provide the actual total well stimulation treatment volume pumped; detail as to each fluid stage pumped, including actual volume, proppant rate or concentration; and actual chemical additive name, type, concentration or rate, and amounts. In lieu of the preceding information, an owner/operator may submit a job log. *055-000-003 Code Wyo. R. §§§ 8, 12, 45 (2012).*

	Pressure monitoring, testing, limitations or other mechanical integrity requirements during well treatment or stimulation
CO	During stimulation, bradenhead[e] annulus pressure shall be continuously monitored and recorded. If at any time during stimulation operations the bradenhead annulus pressure increases more than 200 psi (gauge), the operator shall verbally notify the state as soon as practicable, but no later than 24 hours following the incident. Within 15 days after the occurrence, the operator shall submit a notice giving all details, including corrective actions taken. If intermediate casing has been set on the well being stimulated, the pressure in the annulus between the intermediate casing and the production casing shall also be monitored and recorded. The operator shall keep all well stimulation records and pressure charts on file and available for inspection by the state for at least 5 years. An operator may seek a variance from these bradenhead monitoring, recording, and reporting requirements under appropriate circumstances. *2 Colo. Code Regs. § 404-1(341) (2012).*
ND	• The state may prescribe pretreatment casing pressure testing and other operational requirements to protect wellhead and casing strings during treatment operations. If damage results to the casing or the casing seat from fracturing or chemically treating a well, the operator shall immediately notify the state and proceed with diligence to rectify the damage. If perforating, fracturing, or chemical treating results in irreparable damage that threatens the mechanical integrity of the well, the commission may require the operator to plug the well. • The following mechanical integrity requirements apply to hydraulic fracture stimulations performed through a frac string run inside the intermediate casing string. (1) The frac string must be stung into a liner or run with a packer set at a minimum depth of 100 feet below the top of cement or 100 feet below the top of the Inyan Kara formation, whichever is deeper. (2) The intermediate casing-frac string annulus must be pressurized and monitored during frac operations. (3) An adequately sized, function tested pressure relief valve must be utilized on the treating lines from the pumps to the wellhead, with suitable check valves to limit the volume of flowback fluid should the relief valve open. The relief valve must be set to limit line pressure to no more than 85% of the internal yield pressure of the frac string. (4) An adequately sized, function tested pressure relief valve and an adequately sized diversion line must be utilized to divert flow from the intermediate casing to a pit or containment vessel in case of frac string failure. The relief valve must be set to limit annular pressure to no more than 85% of the lowest internal yield pressure of the intermediate casing string. (5) The surface casing valve must be fully open and connected to a diversion line rigged to a pit or containment vessel. (6) An adequately sized, function tested remote operated frac valve must be utilized between the treating line and the wellhead. • The following specific mechanical integrity requirements apply to hydraulic fracture stimulations performed through an intermediate casing string: (1) The maximum treating pressure shall be no greater than 85% of the American Petroleum Institute rating of the intermediate casing. (2) Casing evaluation tools to verify adequate wall thickness of the intermediate casing shall be run from the wellhead to a depth as close as practicable to 100 feet above the completion formation and a visual inspection with photographs shall be made of the top joint of the intermediate casing and the wellhead flange. If the casing evaluation tool or visual inspection indicates wall thickness is below the American Petroleum Institute minimum or a lighter weight of intermediate casing than the well design called for, calculations must be made to determine the reduced pressure rating. If the reduced pressure rating is less than the anticipated treating pressure, a frac string shall be run inside the intermediate casing. (3) Cement evaluation tools to verify adequate cementing of the intermediate casing shall be run from the wellhead to a depth as close as practicable to 100 feet above the completion formation. If the cement evaluation tool indicates defective casing or cementing a frac string shall be run inside the intermediate casing. If the cement evaluation tool indicates the top of cement behind the intermediate casing is below the top of the Mowry Formation, a frac string shall be run inside the intermediate casing. (4) The intermediate casing and wellhead must be pressure tested to a minimum depth of 100 feet below the top of the Tyler formation for at least 30 minutes with less than 5% loss to a pressure equal or greater than the maximum frac design pressure. (5) If the pressure rating of the wellhead does not exceed the maximum frac design pressure, a wellhead and blowout preventer protection system must be utilized during the frac. (6) An adequately sized, function tested pressure relief valve must be utilized on the treating lines from the pumps to the wellhead, with suitable check valves to limit the volume of flow back fluid should the relief valve open. The relief valve must be set to limit line pressure to no greater than the test pressure of the intermediate casing, less 100 psi; (7) The surface casing valve must be fully open and connected to a diversion line rigged to a pit or containment vessel. (8) An adequately sized, function tested remote operated frac valve must be utilized between the treating line and the wellhead. • If during stimulation, the pressure in the intermediate casing-surface casing annulus exceeds 350 psi, the owner or operator shall verbally notify the state as soon as practicable but no later than 24 hours later. *N.D. Admin. Code 43-02-03-27, 27-1 (2012).*

OH	All casing installed in a well must have a minimum internal yield pressure rating designed to withstand at least 1.2 times the maximum pressure to which the casing may be subjected during stimulation operations. Reconditioned casing that is permanently set in a well shall be hydrostatically pressure tested with an applied pressure at least 1.2 times the maximum internal pressure to which the casing may be subjected or pressure that may be applied during stimulation, whichever is greater, and assuming no external pressure. Test results shall be provided to the state before casing is installed in the well. Wellhead assemblies shall be used to maintain surface control of the well. Each component of the wellhead shall have a working pressure rating equal to or greater than the highest anticipated operating pressure to which the particular component might be exposed during the course of stimulating the well. During stimulation or workover operations, all annuli shall be pressure-monitored. Stimulation or workover operations shall be immediately suspended for any inexplicable pressure deviation above those anticipated increases caused by pressure or thermal transfer. In the event that stimulation fluids circulate, or annular pressures deviate from anticipated, the owner shall immediately notify the state and acquire approval for remediation of casing or cement. If the chief determines that the stimulation of the well has resulted in irreparable damage to the well, the chief shall order that the well be plugged and abandoned within 30 days of issuance of the order. *Ohio Admin. Code Ann. 1501:9-1-08 (2012)*.
PA	No requirements identified in regulations or in statutes.
TX	No requirements identified in regulations or in statutes.
WY	Setting depths of all casing strings shall be determined based on formation fracture gradients and the maximum anticipated pressure to be maintained within the wellbore. *055-000-003 Code Wyo. R. § 22 (2012)*. The Owner/Operator shall provide geological names, geological description and depth of the formation into which well stimulation fluids are to be injected and a detailed description of the proposed well stimulation design, which shall include: the anticipated surface treating pressure range; the maximum injection treating pressure; and the estimated or calculated fracture length and fracture height. The state may require, prior to well stimulation, the owner or operator to perform a suitable mechanical integrity test of the casing or of the casing-tubing annulus or other mechanical integrity test methods. During well stimulation, the Owner/Operator shall monitor and record the annulus pressure at the bradenhead. If intermediate casing has been set on the well being stimulated, the pressure in the annulus between the intermediate casing and the production casing shall also be monitored and recorded. A continuous record of the annulus pressure during the well stimulation shall be submitted to the state. If during the stimulation the annulus pressure increases by more than 500 psig, the Owner or Operator shall verbally notify the Supervisor as soon as practicable but no later than 24 hours following the incident. The Owner or Operator shall include a report containing all details pertaining to the incident, including corrective actions taken. *055-000-003 Code Wyo. R. §§ 22, 45 (2012)*.

Other

CO	An operator making application for approval of an oil and gas location assessment shall provide the surface owner and owners of surface property within 500 feet of the proposed oil and gas location with certain information, including a state information sheet on hydraulic fracturing treatments and must provide notice of subsequent well operations, such as refracturing of a well, at least 7 days in advance of the operations. Placement of all stimulation fluids shall be confined to the objective formations during treatment to the extent practicable. *2 Colo. Code Regs. § 404-1(305, 341) (2012)*.
ND	No additional requirements identified in regulations or in statutes.
OH	• When cementing the production string of a well that will be stimulated by hydraulic fracturing, and the uppermost perforation is less than 500 feet below the base of the deepest underground source of drinking water, sufficient cement shall be used to fill the annular space outside the casing from the seat to the ground surface or to the bottom of the cellar. If it is not so circulated, the owner shall notify the state and perform tests approved by the state. After the top of cement outside the casing is determined, the owner or his/her authorized representative shall contact the state and obtain approval for the procedures to be used to perform any required additional cementing operations. • Using steel production casing with sufficient cement, an oil and gas reservoir shall be isolated during well stimulation and during the productive life of the well. A well shall not be perforated for purposes of well stimulation in any zone that is located around casing that protects underground sources of drinking water without written authorization from the state. • An owner who elects to stimulate a well shall do so in a manner that will not endanger underground sources of drinking water. If during the stimulation of a well, damage to the production casing or cement occurs, and results in the circulation of fluids from the annulus of the surface production casing, the owner shall immediately terminate the stimulation of the well and notify the state. If the state determines that the casing and the cement may be remediated in a manner that isolates the oil and gas bearing zones of the well, the state may authorize the completion of the stimulation of the well. If the state determines that the stimulation of a well resulted in irreparable damage to the well, the state shall order that the

	well be plugged within 30 days. For purposes of determining the integrity of the remediation of the casing or cement of a well that was damaged during the stimulation of the well, the state may require the owner to submit cement evaluation logs, temperature surveys, pressure tests, or a combination of such logs, surveys, and tests. *Ohio Admin. Code Ann. 1501:9-1-08 (2012); Ohio Rev. Code Ann. §§ 1509.17, .19 (2012).*
PA	A well permit application must be accompanied by a plat showing the name of all surface landowners and water purveyors whose water supplies are within 3,000 feet. The applicant must notify the aforementioned landowners and purveyors, and each municipality and storage operator within 3,000 feet. A containment plan for unconventional wells is required. Practices under the plan must be sufficiently impervious and able to contain spilled material until it can be removed or treated, and be compatible with the material to be contained. The plan must be submitted to the state. Containment systems must be used for drilling mud, hydraulic oil, diesel fuel, drilling mud additives, hydraulic fracturing additives, and hydraulic fracturing flowback. Areas where any additives, chemicals, oils, or fuels are to be stored must have sufficient containment capacity to hold the volume of the largest container stored in the area plus 10% to allow for precipitation, unless the container is equipped with individual secondary containment. An owner/operator of a facility conducting natural gas operations in unconventional formations shall submit to the department a source report identifying and quantifying actual air contaminant emissions from any air contamination source. *58 Pa.Cons.Stat. §§ 3211, 3218.2, 3227 (2012).*
TX	No additional requirements identified.
WY	The injection of volatile organic compounds, such as benzene, toluene, ethylbenzene and xylene, (BTEX compounds), or any petroleum distillates into groundwater is prohibited. The proposed use of BTEX compounds or petroleum distillates for well stimulation into hydrocarbon bearing zones is authorized with prior state approval. It is accepted practice to use produced water that may contain small amounts of naturally occurring petroleum distillates, as well stimulation fluid in hydrocarbon bearing zones.
	Following well stimulation, the owner, operator, or service company must provide the actual total well stimulation treatment volume pumped; detail as to each fluid stage pumped, including actual volume, proppant rate or concentration; actual chemical additive name, type, concentration or rate, and amounts; the actual surface pressure and rate at the end of each fluid stage and the actual flush volume, rate and final pump pressure; the instantaneous shut-in pressure, and the actual 15-minute and 30-minute shut-in pressures when available. In lieu of the preceding information, an owner/operator may submit a job log.
	The owner/operator shall provide information to the state as to the amounts, handling and, if necessary, disposal at an identified appropriate disposal facility, or reuse of the well stimulation fluid load recovered during flowback, swabbing, and/or recovery from production facility vessels. Storage of such fluid shall be protective of groundwater as demonstrated by the use of either tanks or lined pits. If lined pits are utilized to store fluid for use in well stimulation, or for reconditioning, for reuse, or to hold for appropriate disposal, then additional requirements to protect wildlife and migratory birds shall be met. *055-000-003 Code Wyo. R. § 45 (2012).*

Source: GAO analysis of state information.

[a]Elements that are viewable on the FracFocus website include date of fracturing treatment, identifying information for wells, and information on hydraulic fracturing fluid composition, including trade name of component, supplier, purpose, ingredients, CAS number, and maximum ingredient concentrations in the additive and in the fluid as a whole.

[b]Pennsylvania law does not specify what information must be disclosed.

[c]29 C.F.R. § 1910.1200(g)(2) (2011).

[d]The disclosures required by this subsection must be made in accordance with the procedures in 29 C.F.R. § 1910.1200(i) (2011) with respect to a written statement of need and confidentiality agreements, as applicable.

[e]A bradenhead is a casing head in an oil well having a stuffing box packed to make a gastight connection.

Table 15: Selected State Requirements—Well Plugging

Requirements for notification, plugging plan or method, witnessing, and reporting

CO	The operator must obtain prior approval of the plugging method. A hole must be plugged so that substances are confined to the reservoir from which they originated. Cement plugs shall be at least 50 feet long and extend at least 50 feet above each zone to be protected. Plugging material, whether cement, a mechanical plug, or other equivalent method approved by the state, must permanently prevent migration of oil, gas, water, or other substance from the formation or horizon in which it originally occurred. Where cement is used, the operator may choose among the following methods of placing cement in the hole: by dump bailer, pumping a balanced cement plug through tubing or drill pipe, pump and plug, or equivalent method approved by the state prior to plugging. Unless prior approval is given, all wellbores shall have water, mud, or other approved fluid between all plugs. The operator must provide notice of the estimated time and date of plugging. Reports of plugging and abandonment must be submitted with a job log or cement verification report from the plugging contractor specifying the fluid used to fill the wellbore, type and slurry volume of cement used, date of work, and depth of plugs. *2 Colo. Code Regs. § 404-1(319) (2012).*
ND	A notice of intention to plug, including the proposed method of plugging and a detailed statement of proposed work, must be approved by the state prior to plugging. Generally, wells must be plugged so to confine permanently all oil, gas, and water in the separate strata originally containing them. This operation shall be accomplished by the use of mud-laden fluid, cement, and plugs, used singly, or in combination, as may be approved by the state. After the plugging of a well, a plugging record shall be filed with the state. *N.D. Admin. Code 43-02-03-33, -34, -31 (2012).*
OH	A permit must be obtained to plug a well. Wells must be plugged so that oil, gas, water, or other fluids shall be confined to the reservoir rock in which it occurs or originates. The owner, or his agent, may have the option of using any method of emplacing the plugging material approved by the state including dump bailer, bullhead, pumping through tubing, casing, or drill pipe. The state may designate an alternate method of plugging in certain areas. Plugging operations must be conducted under the supervision of a state inspector. The owner or his/her agent shall notify the inspector when plugging operations will commence at a dry hole or lost hole in sufficient time to enable the inspector to be present. For all other wells, the owner or agent shall notify the inspector a minimum of 24 hours in advance. The state may grant verbal authorization to commence plugging when the inspector is unable to be present. The state regulations include detail on when plugging must be commenced, when specific types of pipe and casing can be pulled from a well, the materials that can be used for plugging, and how plugging with those materials must proceed. When plugging operations are not witnessed by an inspector, a plugging report on a form provided by the state and signed by the owner or his agent, shall be filed with the state within 30 days after completion of plugging. For all wells plugged with cement, a cementing ticket made by the party cementing the well shall be attached to the plugging report. For all wells plugged with prepared clay, a copy of the prepared clay purchase record shall be attached to the plugging report. When an inspector is present to supervise the plugging operations, a plugging report shall be filed on such form as the chief may prescribe. *Ohio Admin. Code Ann. 1501:9-11-02, -03, -04, -12 (2012).*
PA	Notification and witnessing requirements apply only to wells in coal areas. Prior to plugging a well in an area underlain by a workable coal seam, the operator or owners must notify the state, and the coal operator, lessee or owner to permit representatives to be present at the plugging. Detailed plugging requirements differ based on whether the well is in a coal or noncoal area, whether surface casing is present and how it is attached, and whether the well was stimulated with explosives. A plan is only required if operator proposes an alternate plugging method for approval. Reporting requirements also only apply to coal areas. When plugging of a well in an area underlain by a workable coal seam has been completed, a certification shall be prepared and signed by two experienced and qualified people who participated in the work setting forth the time and manner in which the well was plugged. One copy of the certificate shall be mailed to each coal operator, lessee or owner, and another shall be mailed to the state. *25 Pa. Code §§ 78.91-.98 (2012). 58 Pa.C.S. §§ 3220, 3221 (2012).*
TX	Generally, the operator must give notice of its intention to plug any well prior to plugging. The notice shall set out the proposed plugging procedure as well as the complete casing record. The proposed plugging procedure must be approved before plugging commences. Generally, wells shall be plugged to insure that all formations bearing usable quality water, oil, or gas are protected. Cement plugs shall be set to isolate each productive horizon and usable quality water strata; cement plugs shall be placed by the circulation or squeeze method through tubing or drill pipe. Cement plugs shall be placed by other methods only upon written request with the written approval of the state. The regulations list different specific requirements for wells with surface, intermediate, and production casing, wells with screens or liners, and wells without production casing and open-hold completions. Plugging cannot commence before the date in the notice unless otherwise authorized, and the operator shall notify the district office at least 4 hours before plugging. Exceptions to the timing requirements may be granted in certain circumstances. The operator shall file a plugging record within 30 days of plugging. *16 Tex. Admin. Code 3.14 (2012).*

GAO-12-874 Unconventional Oil and Gas Development

WY	Before plugging, a notice of intent must be filed with the state. The notice must give a detailed statement of proposed work including kind, location, and length of plugs (by depths), and plans for mudding, cementing, shooting, testing, and removing casing, as well as any other pertinent information. Approval must be obtained prior to commencing plugging operations. The regulations list different specific requirements for wells with and without production casing, coalbed methane wells in the Powder River Basin, and wells within a geologic area known as the Special Sodium Drilling Area. When the well has been plugged, a notarized report accompanied by a job log or cement verification report from the plugging contractor specifying the type of fluid used to fill the wellbore, type of slurry volume of cement used, date of work, and depth of plugs placed must be submitted. *055-000-003 Code Wyo. R. §§ 15, 18 (2012).*

Source: GAO analysis of state information.

Table 16: Selected State Requirements—Site Reclamation

Backfilling, regrading, recontouring, and compaction alleviation requirements

CO	Areas disturbed by drilling and subsequent operations no longer needed for production will be restored to their original condition or their final land use as designated by the surface owner and shall be maintained to control dust and minimize erosion. If subsidence occurs on crop lands, additional topsoil shall be added, and the land shall be releveled as close to its original contour as practicable. Interim reclamation must occur no later than 3 months on crop land or 6 months on noncrop land after operations cease unless that time is extended by the state. Areas needed for subsequent operations within the year shall be stabilized and maintained to minimize dust and erosion. Areas compacted by drilling and subsequent operations no longer needed for production shall be cross-ripped. On crop land, operations shall be undertaken when soil moisture is below 35% of field capacity to a depth of 18 inches unless bedrock is shallower. After well plugging, all access roads to plugged wells shall be closed, graded, and recontoured. Culverts and other obstructions shall be removed. As applicable, compaction alleviation will be performed to the same standards as for interim reclamation. Final reclamation must be completed within 3 months on crop land and 12 months on noncrop land unless that time is extended by the state. Reclamation of the well site and access road shall be considered complete on crop land when, among other things, observation over two growing seasons indicates no significant unrestored subsidence. Stabilization so as to minimize erosion to the extent practicable is a factor in determining completion of final reclamation for all disturbed areas. *2 Colo. Code Regs. § 404-1(1003, 1004) (2012).*
ND	The well site, access road, and other associated facilities constructed for the well shall be reclaimed within a year after a well is plugged or a permit expires or is canceled or revoked. Operators must submit and obtain approval of a reclamation plan, including a description of the proposed work, including topsoil redistribution and reclamation plans for the access road and other associated facilities. Gravel and other surfacing material must be removed, stabilized soil must be remediated, and the well site, access road, and other associated facilities shall be reshaped as near as practicable to their original contour. Previously stockpiled topsoil shall be evenly distributed over the disturbed area. *N.D. Admin. Code 43-02-03-34.1 (2012).*
OH	Unless the state approves a longer time period, within 3 months after drilling commences in an urbanized area and within 6 months after drilling commences in all other areas, the owner or the owner's agent shall grade or terrace disturbed areas that are not required in production. Within 3 months after a well is plugged in an urbanized area, and within 6 months after a well is plugged in all other areas, or after the plugging of a dry hole, unless the state approves a longer time period, the owner or the owner's agent shall fill remaining excavations. *Ohio Rev. Code Ann. § 1509.072 (2012).*
PA	Each oil or gas well owner or operator shall restore the land surface within the area disturbed in siting, drilling, completing, and producing the well. Within 9 months after completion of drilling a well, the owner or operator shall restore the well site, remove or fill all pits used to contain produced fluids or industrial wastes, remove all production and storage facilities, supplies and equipment, and remove all drilling supplies and equipment not needed for production. Drilling supplies and equipment not needed for production may be stored on the well site if express written consent is obtained from the surface landowner. This time frame may be extended to a maximum of 2 years upon request and submission of a plan demonstrating that the extension will result in less earth disturbance or that site restoration cannot be achieved due to adverse weather conditions or a lack of essential fuel, equipment or labor. *58 Pa.C.S. § 3216.*

TX	The operator shall fill the rathole, mouse hole, and cellar, and shall empty all tanks, vessels, related piping, and flowlines that will not be actively used in the continuing operation of the lease within 120 days after plugging work is completed. Within the same 120-day period, the operator shall remove all such tanks, vessels, and related piping, remove all loose junk and trash from the location, and contour the location to discourage pooling of surface water at or around the facility site. The operator shall close all pits. The district director or the director's delegate may grant a reasonable extension of time of not more than an additional 120 days for the removal of tanks, vessels and related piping. *16 Tex. Admin. Code 3.14 (2012).*
WY	Reclamation must be initiated within 1 year of permanent abandonment of a well or last use of a pit and must be completed in accordance with the landowner's reasonable requests and/or resemble the original contour of adjoining lands. All disturbed areas on state lands will be recontoured unless the state approves otherwise. *055-000-003 Code Wyo. R. §§ 7, 17 (2012).*

Revegetation requirements

CO	When a well is completed for production, all disturbed areas no longer needed will be revegetated as soon as practicable. For crop lands, all segregated soil horizons shall be replaced to their original relative positions and contour and shall be tilled to establish a proper seedbed. The area shall be treated if necessary and practicable to prevent weeds and erosion. Previously present perennial forage crops shall be reestablished. For noncrop lands, all segregated soil horizons shall be replaced to their original relative positions and contour as near as practicable to achieve erosion control and long-term stability and shall be tilled to establish a proper seed bed. The area shall be reseeded in the first favorable season following rig demobilization. Reseeding consistent with adjacent plant communities is encouraged. Seed mix should be as agreed with surface owner or based on consultation with the local soil conservation district. To be considered complete, reclamation must, among other things, establish uniform vegetative cover reflecting predisturbance or reference area vegetation with total cover of at least 80% of predisturbance levels, excluding weeds. After a well is plugged, revegetation of well sites, associated production facilities, and access roads shall be performed to the same standards. A decision as to whether final reclamation has been completed will take into account permanent physical erosion reduction methods as an alternative to 80% revegetation. *2 Colo. Code Regs. § 404-1(1003, 1004) (2012).*
ND	The reclamation plan for the well site, access road, and other associated facilities shall include a reseeding plan, if applicable. Disturbed areas shall be revegetated with native species or according to the reasonable specifications of the government land manager or surface owner. N.D. Admin. Code 43-02-03-34.1 (2012).
OH	Unless the state approves a longer time period, within 3 months after the date upon which the surface drilling of a well is commenced in an urbanized area, and within 6 months after the date upon which the surface drilling of a well is commenced in all other areas, the owner or the owner's agent shall plant, seed, or sod the area disturbed that is not required in production of the well where necessary to bind the soil and prevent substantial erosion and sedimentation. Within 3 months after a well that has produced oil or gas is plugged in an urbanized area, and within 6 months after a well that has produced oil or gas is plugged in all other areas, or after the plugging of a dry hole, unless the chief approves a longer time period, the owner or the owner's agent shall plant, seed, or sod the area disturbed where necessary to bind the soil and prevent substantial erosion and sedimentation. *Ohio Rev. Code Ann. § 1509.072 (2012).*
PA	Upon final completion of an earth disturbance activity or any stage or phase of activity, the site must have topsoil immediately restored, replaced, or amended, seeded, mulched, or otherwise permanently stabilized and protected from accelerated erosion and sedimentation. For the earth disturbance activity or any stage or phase of an activity to be considered permanently stabilized, the disturbed area must be covered with a minimum uniform 70% perennial vegetative cover with a density capable of resisting accelerated erosion and sedimentation or an acceptable best management practice that permanently minimizes accelerated erosion and sedimentation.
	See also specific pit closure requirements above. In addition, where residual waste is disposed of by land application, the application area shall be revegetated to stabilize the soil surface. The revegetation shall establish a diverse, effective permanent vegetative cover which is capable of self-regeneration and plant succession. Where vegetation would interfere with the intended use of the surface by the landowner, the surface shall be stabilized against erosion. *25 Pa Code §§ 78.63, 102.22 (2012).*
TX	No requirements identified in regulations or in statutes.[a]
WY	Reclamation must be initiated within 1 year of permanent abandonment of a well or last use of a pit and must be completed in accordance with the landowner's reasonable requests and/or resemble the original vegetation of adjoining lands. All disturbed areas on state lands will be reseeded unless the state approves otherwise. *055-000-003 Code Wyo. R. §§ 7, 17 (2012).*

Source: GAO analysis of state information.

[a]According to Texas state officials, however, requirements may be included in permits issued by the Railroad Commission.

Table 17: Selected State Requirements—Waste Management in Pits

Pit siting requirements (with regard to sensitive areas)

CO	Generally, pits shall not be constructed in areas where pathways for communication with groundwater or surface water are likely to exist. Operations at new locations within the intermediate and external buffer zones surrounding a public water system cannot use pits.[a] 2 Colo. Code Regs. § 404-1(317B, 902) (2012).
ND	Reserve pits may be used for wells drilled to certain depths providing the pit can be constructed, used, and reclaimed in a manner that will prevent pollution of the land surface and freshwaters. In special circumstances, based on site-specific conditions, the state may prohibit construction of a reserve pit or may impose more stringent pit construction and reclamation requirements. Drilling pits and reserve pits shall not be located in, or hazardously near, bodies of water, nor shall they block natural drainages. No pit shall be wholly or partially constructed in fill dirt unless approved by the state. *N.D. Admin. Code 43-02-03-19.4, 19.5 (2012).*
OH	Drilling permits for urbanized areas are conditioned on the state receiving direct notification at least 48 hours prior to pit construction. All pits used for temporary storage of saltwater and oil field wastes shall not be used in an area that is subject to flooding by streams, rivers, lakes, or drainage ditches, unless so constructed that the pits would not normally be affected by flooding. *Ohio Admin. Code Ann. 1501:9-1-02,-3-08 (2012).*
PA	Generally, pits for the control, storage, and disposal of production fluids may not be located within 100 feet of a stream, body of water, or wetland. Pits for the disposal of drill cuttings may not be located within 100 feet of a stream, body of water, or wetland unless otherwise permitted, and may not be located within 200 feet of a water supply. Pits for the disposal of residual waste may not be located within 100 feet of a stream, body of water, or wetland or within 200 feet of a water supply. Generally, pit bottoms must be at least 20 inches above the seasonal high groundwater table. *25 Pa. Code § 78.56, 57, .61, .62 (2012).*
TX	No commercial oil and gas waste disposal pits may be constructed in any coastal natural resource area[b] and all oil and gas waste disposal pits shall be designed to prevent releases of pollutants that adversely affect coastal waters or critical areas. *16 Tex. Admin. Code § 3.8 (2012).*[c]
WY	Owners or operators must obtain state approval for the location of noncommercial centralized pits, reserve pits, and workover and completion and produced water pits proposed for critical areas. Pits in critical areas include those located within ¼ mile of water supplies, areas where groundwater is less than 20 feet from the surface, locations which are within 500 feet of wetlands, ponds, lakes, perennial drainages or within a floodplain, and areas where pit fluids are greater than 10,000 mg/l total dissolved solids. When a retaining pit is located in an area with a high potential for communication between pit contents and surface water or shallow groundwater, or to protect people, livestock, or wildlife, the state may require changes to plans including running a closed system, lining the pit, or installing monitoring systems and providing additional reporting. In areas where groundwater is less than 20 feet below the surface, a closed system must be utilized. Generally, pits cannot be located closer than 350 feet from water supplies. *055-000-001 Code Wyo. R. § 2; 055-000-003 Code Wyo. R. § 22; 055-000-004 Code Wyo. R. § 1 (2012).*

Pit lining requirements

CO	Certain pits, including drilling pits for fluids containing hydrocarbon or chloride concentrations exceeding certain levels; production pits in certain regions unless the quality of the produced water is as good or better than that of the underlying groundwater or seepage will not reach the underlying aquifer or waters of the state at levels in excess of applicable standards; special purpose pits excluding emergency pits and certain flare pits; skim pits; and multiwell pits in certain regions used to contain produced water, drilling fluids, or completion fluids that will be recycled or reused, must be lined if they were constructed after Spring of 2009. Liners shall be synthetic, impervious, have high puncture and tear strength and adequate elongation, and be resistant to ultraviolet light deterioration, weathering, hydrocarbons, acids, alkali, fungi and other substances in produced water. All pit lining systems shall be designed, constructed, installed, and maintained in accordance with the manufacturers' specifications and good engineering practices. Field seams must be installed and tested in accordance with manufacturer specifications and good engineering practices. Unless an operator can demonstrate equivalent protection with an alternative system, liners for on-site pits must also meet the following requirements. Liners shall have a minimum thickness of 24 mils and must cover bottom and insides of pit with enough overhang to be secured in a 12-inch anchor trench. Foundation for the liner shall be constructed with (1) at least 12 inches of soil compacted such that the amount of liquid it can conduct does not exceed 1.0×10^{-7} centimeters per second or (2) with other material if two liners of at least 24 mils in thickness are used and the pit bottom and sides are padded and free of material that could puncture the liner. In Sensitive Areas, a leak detection system, increased record-keeping, monitoring, or underlying gravel fill sumps and lateral systems may be required. *2 Colo. Code Regs. § 404-1(904) (2012).*

ND	Generally, no saltwater, drilling mud, crude oil, waste oil, or other waste shall be stored in earthen pits or open receptacles except in an emergency and upon state approval. A lined earthen pit or open receptacle may be temporarily used to retain oil, water, cement, solids, or fluids generated in well completion, servicing, or plugging. Such a pit or receptacle must be sufficiently impermeable to provide adequate temporary containment. Pit contents must be removed within 72 hours after operations have ceased and disposed of at an authorized facility. Freshwater pits must also be lined. *N.D. Admin. Code 43-02-03-19.3 (2012).*
OH	No requirements identified in regulations or in statutes, though pits used for the temporary storage of saltwater and oil field wastes shall be "liquid tight." *Ohio Admin. Code Ann. 1501:9-3-08 (2012).*
PA	Pits used for temporary containment, for control, storage, and disposal of production fluids, and for disposal of residual waste must be lined. Liners must meet a specific permeability threshold and be of sufficient strength and thickness to maintain the integrity of the liner. For pits other than those used for temporary containment, the minimum liner strength must be 30 mils. Liners shall be sealed together to prevent leakage in accordance with the manufacturer's directions. The liner shall be designed, constructed, and maintained so that the physical and chemical characteristics of the liner are not adversely affected by the waste and the liner is resistant to physical, chemical, and other failure during transportation, handling, installation, and use. Alternate liners or materials may be approved. Pits must be smooth so they do not tear the liner and must be able to bear the weight of their contents without settling that may affect the liner integrity. If the pit bottom or sides consist of rock, shale, or other materials that may cause the liner to fail, a subbase of at least 6 inches of soil, sand, or smooth gravel, or sufficient amount of an equivalent material, shall be installed over the area as the subbase for the liner. *25 Pa. Code §§ 78.56, .57, .62 (2012).*
TX	A permit issued to maintain or use any lined pit for storage or disposal of oil field brines or other mineralized waters will contain requirements relating to liner material, thickness, procedures for installing liners, schedules for inspecting and/or replacing liners. A permit issued to maintain or use a pit for storage of oil field fluids or oil and gas wastes may contain such requirements. *16 Tex. Admin. Code § 3.8 (2012).*
WY	Before drilling commences, approval to construct reserve pits must be applied for and received. Special precautions, including an impermeable liner and/or membrane, shall be taken if necessary to prevent water contamination and where drilling is conducted close to water supplies, residences, schools, hospitals, or other structures. Unlined pits shall not be constructed in fill. Lining of pits with reinforced oilfield grade material, compatible with the waste to be received, will be required under certain circumstances including pits proposed to be constructed in critical areas as well as on sites with sandy soils, shallow groundwater, in groundwater recharge areas, or sites immediately adjacent to the Green River or the Colorado River drainage and other sensitive environments or circumstances. Pits constructed in fill or those used to retain oil base drilling muds, high-density brines, and/or completion or treating fluids must be lined. Pits constructed to retain produced water with a total dissolved solids concentration in excess of 10,000 mg/l must be lined. Pits retaining water with a total dissolved solids concentration less than 10,000 mg/l may be required to be lined on a case-by-case basis. Soil mixture liners, recompacted clay liners, and manufactured liners must be compatible with the waste contained. Synthetic liners must meet the following specifications: a 9 to 12 mil thickness, greater than 20% elongation at failure, puncture strength of 60 lbs, tear strength of 50 lbs, and permeability less than 10-7 cm/sec. Joints must be overlapped at least 2 inches and seams sealed per manufacturer recommendation. Blemishes, holes, or scars must be repaired per manufacturer recommendation. Breaches for equipment must be reinforced. Slopes shall not exceed 3:1 for soil mixture or recompacted liners or 1:1 for manufactured liners. Reasonable provisions for protection of liners during filling and emptying activities must be included in the construction plans. Manufactured liners must be installed over smooth fill that is free of pockets or materials which could damage the liner. Sand, sifted dirt ,or bentonite are suggested. At no time will straw or any other organic material except synthetic cushion fabric designed for that purpose be used for a liner cushion. Installation of synthetic or soil mixture liners must be in accordance with accepted engineering practice. Liner edges must be secured by placing in a trench which is deep enough to receive approximately 1' of compacted soil which will anchor the material. *055-000-003 Code Wyo. R. § 22, 055-000-004 Code Wyo. R. § 1 (2012).*

Freeboard[d] and secondary containment requirements for pits and tanks

CO	•	Pits shall be constructed, monitored, and operated to provide for a minimum of 2 feet of freeboard at all times between the top of the pit wall at its point of lowest elevation and the fluid level of the pit. A method of monitoring and maintaining freeboard shall be employed.
	•	For new operations at new locations within the intermediate buffer zone surrounding a public water system and certain operations that create a new surface disturbance but were otherwise in existence before the spring of 2009 and are in the internal buffer zone surrounding a public water system, flowback and stimulation fluids must be contained within tanks that are either on the well pad or in an area with downgradient perimeter berming. Berms or other containment

	devices must be constructed around crude oil, condensate, and produced water storage tanks. Tanks containing oil, condensate, or produced water with greater than 3,500 mg/l total dissolved solids shall have secondary containment sufficient to hold the contents of the largest single tank and enough freeboard for precipitation. *2 Colo. Code Regs. §§ 404-1(317B, 604, 902, 906) (2012).*
ND	When necessary to prevent pollution of the land surface and freshwaters, the state may require the drill site to be sloped and diked. Pits containing drill cuttings and other solids must be diked so as to prevent surface water from running into the pit. Dikes must be erected and maintained around oil tanks at any production facility and saltwater tanks at any saltwater handling facility built or rebuilt on or after July 1, 2000, within 30 days after the well has been completed. Dikes must be erected and maintained around tanks and facilities built earlier when deemed necessary by the state. Dikes as well as the base material under the dikes and within the diked area must be constructed of sufficiently impermeable material to provide emergency containment. Dikes must be of sufficient dimension to contain the total capacity of the largest tank plus 1 day's fluid production. The required capacity of the dike may be lowered if need can be demonstrated. Discharged saltwater liquids or brines must be properly removed and may not be allowed to remain standing within or outside of any diked areas. *N.D. Admin. Code 43-02-03-19, -19.4, -49, -53 (2012).*
OH	All pits shall have a continuous embankment surrounding them sufficiently above the level of the surface to prevent surface water from entering. In order to protect life, health, and property the state may require where a clear and present hazard exists that any producing equipment at the well-head and related storage tanks be protected by an earthen dike or earthen pit that shall have a capacity sufficient to contain any substances resulting, obtained, or produced in connection with the operation of the related oil or gas well. The dike or pit shall be maintained for the purpose for which it was constructed, and the reservoir within shall be kept reasonably free of water and oil. *Ohio Admin. Code Ann. 1501:9-3-08, 1501:9-9-05 (2012).*
PA	• Pits used for temporary containment or the control, storage, and disposal of production fluids must maintain 2 feet of freeboard at all times. If open tanks are used for temporary containment, 2 feet of freeboard must remain at all times unless the tank is provided with an overflow system with sufficient volume. If an open standby tank is used, it shall be maintained with 2 feet of freeboard. If this requirement is violated, the operator immediately shall take the necessary measures to ensure the structural stability of the pit or tank, prevent spills, and restore the 2 feet of freeboard. • See above for secondary containment provisions applicable specifically to unconventional wells. In addition, if an owner or operator uses a tank with a capacity of at least 660 gallons or tanks with a combined capacity of at least 1,320 gallons to contain oil produced from a well, the owner or operator shall construct and maintain a dike or other method of secondary containment that satisfies the requirements of certain federal rules[e] around the tank or tanks which will prevent the tank contents from entering waters of this Commonwealth. The containment area shall have capacity sufficient to hold the volume of the largest single tank, plus a reasonable allowance for precipitation based on local weather conditions and facility operation. *25 Pa. Code §§ 78.56, .57, .64 (2012).*
TX	A permit to maintain or use a pit for storage of oil field fluids or oil and gas wastes may contain requirements including dike design, overflow warning devices, and leak detection devices. *16 Tex. Admin. Code 3.8 (2012).*
WY	Operators are reminded to comply with federal regulations[f] that require facilities to construct appropriate containment or diversionary structures or equipment to prevent discharged oil from reaching waters of the United States. Liquids in pits must be kept at a level that takes into account extreme precipitation events and prevents overtopping and unpermitted discharges. *055-000-004 Code Wyo. R. §§ 1, 4 (2012).*

Pit closure requirements

CO	All pits unnecessary for further operations, excluding the drilling pit, must be backfilled as soon as possible after the drilling rig is released to conform with surrounding terrain. Drill pits must be closed after drilling and completion activities conclude: no more than 3 months later for crop land, and no more than 6 months for noncrop land. Drilling fluids must be removed from drill pits and soils must meet contaminant concentration levels specified in the rules. Material removed from the pit for drying shall be returned prior to backfilling, and only de minimis amounts may be incorporated into surface material. Dry pits shall be backfilled to return soils to their original relative positions. Subsidence within 2 years must be corrected. On crop land, or within the 100-year floodplain, reclamation shall not form an impermeable barrier in the pit and at least 3 feet of backfill shall be applied over any remaining drilling pit contents. Emergency pits shall be closed and remediated as soon as the initial phase of emergency response operations are complete or process upset conditions are controlled. Upon plugging of a well, all other pits must be backfilled. Pits other than drilling pits must be closed in accordance with a remediation plan approved by the state. General site investigation and remediation requirements include a sensitive area determination, sampling and analysis of soil and groundwater to determine the extent of any contamination, removal and management of exploration and production waste, and remediation of contaminated soil and groundwater. Synthetic liners must be removed and disposed of. Constructed soil liner material may be removed for treatment or disposal, or ripped and mixed with native soils such that it

	continues to meet applicable soil concentration levels. Pits must be backfilled to return the soils to their original relative positions. If there is subsidence, additional topsoil shall be added and the land shall be releveled as close to its original contour as practicable. *2 Colo. Code Regs. § 404-1 (1003, 905, 909, 1004) (2012).*
ND	A lined earthen pit or open receptacle may be temporarily used to retain oil, water, cement, solids, or fluids generated in well completion, servicing, or plugging operations. The contents of earthen pits or open receptacles must be removed within 72 hours after operations have ceased and must be disposed of at an authorized facility. Pits must be reclaimed and open receptacles must be removed within 30 days after operations have ceased. Drill cuttings and solids generated during well drilling and completion may be buried in pits provided that the pit can be reclaimed in a way that will prevent pollution of the land surface and freshwaters. Drilling pits must be reclaimed within 30 days after drilling or the expiration of a drilling permit. Reserve pits can be used in certain circumstances to contain certain solids and fluids used and generated during well drilling and completion operations, provided that the pit can be reclaimed in a manner that will prevent pollution of the land surface and freshwaters. Reserve pits must be reclaimed within a year of the completion of a shallow well or prior to drilling below the surface casing shoe on any other well. Prior to reclaiming a pit, approval of a pit reclamation plan must be obtained from the state. Any water or oil accumulated on the pit must be removed prior to reclamation. Drilling waste shall be encapsulated in the pit and covered with at least 4 feet of backfill and topsoil and surface sloped, when practicable, to promote surface drainage away from the reclaimed pit area. In certain circumstances, the state may impose more stringent reclamation requirements for pits. *N.D. Admin. Code 43-02-03-19.3, -19.4, -19.5 (2012).*
OH	Each drilling permit issued in an urbanized area will be conditioned on the state receiving direct notification a minimum of 48 hours prior to pit closure. Pits may be used for the temporary storage of saltwater and oil field wastes but no pit may be used for the ultimate disposal of saltwater. Saltwater and oil field wastes must be drained or removed and properly disposed of periodically, at intervals not to exceed 180 days. Pits may be used for the temporary storage of frac-water and other liquid substances produced from the fracturing process, but upon termination of the fracturing process, pits not otherwise permitted shall be emptied, the contents disposed of and the pits filled in, unless this requirement is waived or extended. Within 14 days after the date upon which the drilling of a well is completed to total depth in an urbanized area and within 2 months after the date upon which the drilling is completed in all other areas, the owner or his agent, in accordance with a restoration plan filed with the state, must fill all the pits for containing brine and other waste substances resulting, obtained, or produced in connection with exploration or drilling for oil or gas that are not required by other state or federal law or regulation. *Ohio Admin. Code Ann. 1501:9-1-02, -3-08 (2012); Ohio Rev. Code Ann. § 1509.072 (2012).*
PA	Generally, pits used for temporary containment must be removed or filled within 9 months after completion of drilling or within 90 days of construction for pits used during servicing, plugging, and recompleting. Upon abandonment of a well, the operator shall restore pits used to store production fluids by removing and disposing of the contents of the pit, including the liner. The pit shall be backfilled to the ground surface. Dewatered, uncontaminated drill cuttings may be buried at the site where they were generated in structurally sound pits. The pit must be backfilled to the ground surface. Residual waste,[9] including contaminated drill cuttings, may be buried at the site where it was generated in structurally sound, impermeable, lined pits. Free liquid must be removed from the pit prior to waste encapsulation and a liner must cover the contents so that water does not infiltrate. The pit shall be backfilled to at least 18 inches over the top of the liner. Residual waste may not be disposed of at the well site if it exceeds specified concentrations. In all cases, backfilled pits must be graded to promote runoff with no depressions. The stability of backfilled pits must be compatible with the adjacent land. The surface of the backfilled pit must be revegetated or otherwise stabilized against accelerated erosion consistent with land use. *25 Pa. Code §§ 78.56, .57, .61, .62 (2012).*
TX	A person who maintains or uses a reserve pit, mud circulation pit, fresh makeup water pit, fresh mining water pit, completion/workover pit, basic sediment pit, flare pit, or water condensate pit shall dewater, backfill, and compact the pit. Reserve pits and mud circulation pits must be closed within a year of the completion of drilling operations; if they contain fluids exceeding a certain chloride concentration, they must be dewatered within 30 days. All completion/workover pits shall be dewatered within 30 days and closed within 120 days of well completion or workover. Basic sediment pits, flare pits, freshwater pits, and water condensate pits shall be dewatered and closed within 120 days of cessation of use. The state may require that pits be backfilled sooner if oil and gas wastes or oil field fluids are likely to escape from the pit or the pit is being used for improper storage or disposal of oil and gas wastes or oil field fluids. Prior to backfilling, all oil and gas wastes which are in the pit must be disposed of. *16 Tex. Admin. Code 3.8 (2012).*

WY	If the pit is proposed to be closed through the usual method of on-site natural evaporation and subsequent burial of solids, if pit treatment procedures are going to be applied, or if closure plans have changed from the original proposal, or any time wastes are disposed off-site, a notice must be submitted and approved prior to closure. Oil and Gas Conservation Commission staff must be provided the opportunity to witness closure. Verbal notice of at least 24 hours prior to closure is required. Oil, water, and other fluids must be immediately removed from temporary emergency pits and disposed. Trenching or squeezing pits is expressly prohibited. Burial methods cannot compromise the integrity of manufactured, soil mixture, or recompacted clay liners without written approval. Closure standards and testing requirements for all pits will be determined by the state based on site-specific conditions. Pit solids showing high concentrations of salt must be removed from the location and disposed in a permitted facility, encapsulated, or chemically or mechanically treated. When drilling with oil-based muds, oil-based mud solids must be removed and disposed in a permitted facility; solidified using a commission-approved commercial pit treatment, roadspread, landspread, landfarmed; or, bioremediated. Burial after encapsulation or solidification will be approved if the stabilized mixture contains less than 10 mg/l leachable oil and less than 5,000 mg/l leachable dissolved solids. Reserve pits containing oil, sheens, condensate, other hydrocarbons or chemicals proven to be hazardous shall undergo fluid removal as soon as practical or shall be fenced and netted to avoid loss of animals and birds. The state may require testing of wastes and additional disposal requirements prior to closure of a pit if it has reason to believe exempt exploration and production wastes have been commingled with hazardous wastes. An Operator or Owner wishing to treat pits for closure must submit to the commission a plan outlining the objectives that the treatment is designed to achieve. Production pit areas and reserve pits will be reclaimed after they have dried sufficiently following the removal of any oil, sheens, or other hydrocarbons, or if they contain hazardous chemicals. Pits used solely for water retention in coalbed methane areas in the Powder River Basin may be left open with state approval at the landowner's request. *055-000-004 Code Wyo. R. § 1 (2012).*

Source: GAO analysis of state information.

[a]Generally, operations may not occur at all within the internal buffer zone surrounding a public water system, so pits may not be located there either.

[b]A coastal natural resource areas include coastal barriers, coastal historic areas, coastal preserves, coastal shore areas, coastal wetlands, critical dune areas, critical erosion areas, gulf beaches, hard substrate reefs, oyster reefs, submerged lands, special hazard areas, submerged aquatic vegetation, tidal sand or mud flats, water in the open Gulf of Mexico, or water under tidal influence, as these terms are defined in Texas law.

[c]In addition, according to Texas state officials, permit applications for waste management are evaluated to determine proximity to sensitive areas and may be denied if the proposed facility is to be located in or near a sensitive area. "Sensitive areas" are "defined by the presence of factors, whether one or more, that make an area vulnerable to pollution from crude oil spills. Factors that are characteristic of sensitive areas include the presence of shallow groundwater or pathways for communication with deeper groundwater; proximity to surface water, including lakes, rivers, streams, dry or flowing creeks, irrigation canals, stock tanks, and wetlands; proximity to natural wildlife refuges or parks; or proximity to commercial or residential areas." 16 Tex. Admin. Code § 3.91 (2012).

[d]Freeboard is the height that is above the recorded highwater mark of a structure associated with a body of water and that is an allowance against overtopping by waves or other transient disturbances.

[e]40 C.F.R. pt. 112 (2011) (relating to oil pollution prevention).

[f]40 C.F.R. pt. 112 (2011) (relating to oil pollution prevention).

[g]"Residual waste" means any garbage, refuse, other discarded material or other waste including solid, liquid, semisolid, or contained gaseous materials resulting from industrial, mining and agricultural operations and any sludge from an industrial, mining or agricultural water supply treatment facility, waste water treatment facility or air pollution control facility, provided that it is not hazardous.

Table 18: Selected State Requirements—Waste Management through Underground Injection[a]

Requirements regarding existing wells

CO	Application for a well shall include (1) plan of the area within 1/4 mile of the proposed disposal well showing the location of all oil and gas wells, domestic and irrigation wells of public record; and (2) the identification of all oil and gas wells currently producing from the proposed injection zone within 1/2 mile of the disposal zone. Remedial action shall be required for any well within one-quarter (1/4) mile of the proposed disposal well in which the injection zone is not adequately confined. The application must identify the need for such remedial action and a plan for the performance of such work. *2 Colo. Code Regs. § 404-1 (325) (2012).*
ND	Applications must include a plat of the area of review (1/4-mile radius) and detailing the location, well name, and operator of all wells in the area of review. The plat should include all injection wells, producing wells, plugged wells, abandoned wells, drilling wells, dry holes, and water wells. The application is also to identify the need for corrective action on wells penetrating the injection zone in the area of review. Before injection commences in an underground injection well, the applicant must complete any needed corrective action on wells penetrating the injection zone in the area of review. *N.D. Admin. Code 43-02-05-04 (2012).*
OH	Application to include a map showing the geographic location of all wells penetrating the formation proposed for injection, regardless of status, within the area of review (1/4 mile to 1/2 mile from the well depending on its volume). Application must also include a proposed corrective action of wells penetrating the proposed injection formation or zone within the area of review, if required to ensure the injection well will not cause or allow movement of fluid into a source of underground water. *Ohio Admin. Code Ann. 1501:9-3-06, 9-3-12 (2012).*[b]
PA[c]	Operator must identify the location of all known wells in the area of review that penetrate the injection zone (or for wells operating over the fracture pressure of the injection formation, all known wells in the area of review penetrating formations affected by the increase in pressure). For such wells that are improperly sealed, completed, or abandoned, the operator shall also submit a plan of actions necessary to prevent movement of fluid into underground sources of drinking water ("corrective action"); status of corrective actions is considered in permit review and a compliance schedule may be a permit condition.[d,e] *40 C.F.R. §§ 144.31, 144.55, 146.7, 146.24 (2012).*
TX	Applicants shall, based on review of the public record, identify wells that penetrate the proposed disposal zone within a 1/4 mile radius of the proposed disposal well to determine if all abandoned wells have been plugged in a manner that will prevent the movement of fluids from the disposal zone into freshwater strata, and identify any wells which appear to be unplugged or improperly plugged and any other such wells of which the applicant has actual knowledge; unless a variance is granted. (No specific regulatory provision for corrective action.[f]) *16 Texas Admin. Code § 3.9 (2012).*
WY	Application to include plan showing the location of the disposal well or wells, including abandoned and drilling wells and dry holes; and investigation of mechanical conditions of all wells which have penetrated the disposal zone within 1/4 mile radius of the proposed disposal well. *Wyo. Code R. 055-000-004 § 5 (2012).*

Casing/cementing

CO	No specific requirements for disposal wells, but applications must include information on casing and cement bond log.[g,h] (See also integrity testing.) *2 Colo. Code Regs. § 404-1 (325) (2012).*
ND	All injection wells shall be cased and cemented to prevent movement of fluids into or between underground sources of drinking water or into an unauthorized zone. The casing and cement used in construction of each new injection well shall be designed for the life expectancy of the well. In determining and specifying casing and cementing requirements, all of the following factors shall be considered: • depth to the injection zone, • depth to the bottom of all underground sources of drinking water, • estimated maximum and average injection pressures, • fluid pressure, • estimated fracture pressure, and • physical and chemical characteristics of the injection zone. *N.D. Admin. Code 43-02-05-06 (2012).*

OH	Surface casing shall be free of apparent defects and set at least 50 feet below the deepest underground source of water containing less than 10,000 mg/L chlorides, and sealed by circulating cement to the surface under the supervision of the state. Casing to be mechanically centralized and enclosed in cement to a height no less than 300 feet above the top of the injection zone. The cement bond log or cement records are to be submitted, or the state is to verify the number of sacks of cement. State inspector to be notified in advance. *Ohio Admin. Code Ann. 1501:9-3-05 (2012).*
PA	Wells shall be cased and cemented to prevent movement of fluids into or between protected aquifers.[i] Surface casing shall be installed and cemented from the surface to at least 50 feet below the base of the lowermost protected aquifer, and for brine disposal wells, install long string casing and tubing extending to the injection zone and cement to a point 50 feet above the injection zone. Design shall consider the depth to injection zone, depth to the bottom of the aquifer, and the estimated injection pressures. *40 C.F.R. §§ 146.22,147.1955 (2012).*
TX	Disposal wells shall be cased and the casing cemented in such a manner that the injected fluids will not endanger oil, gas, geothermal resources, or freshwater resources. Disposal wells must meet casing and cementing requirements for production wells, such as: • use of pressure-tested steel casing; • anchoring of casing; • all usable-quality water zones must be isolated and sealed off to effectively prevent contamination or harm; • requirements for surface, intermediate, and production casing; • cementing by the pump and plug method; and • pressure-test standards during cementing. *16 Texas Admin. Code §§ 3.9, .13 (2012).*
WY	Disposal wells shall be cased and the casing cemented in such a manner that damage will not be caused to oil, gas, or freshwater sources. The disposal application shall include a description of the casing in the disposal well or wells, or the proposed casing program and the proposed method for testing casing before use of the disposal well or wells. *Wyo. Code R. 055-000-004 § 5 (2012).*

Operating pressure requirements

CO	Operator to indicate operating pressures on application. Maximum injection pressure will be set by the Director upon approval. *2 Colo. Code Regs. § 404-1 (325) (2012).*
ND	Injection pressure at the wellhead shall not exceed a maximum that shall be calculated so as to assure that the pressure in the injection zone during injection does not initiate new fracture or propagate existing fractures in the confining zone adjacent to the freshwater resource. In no case shall injection pressure initiate fractures in the confining zone or cause the movement of injection or formation fluids into an underground source of drinking water. *N.D. Admin. Code 43-02-05-09 (2012).*
OH	The maximum allowable operating pressure for any injection well shall be determined by a formula or method approved by the state. Under no circumstances shall liquids or waste matter from any source, other than saltwater from oil and gas operations or standard well treatment fluid, be injected into any injection well. *Ohio Admin. Code Ann. 1501:9-3-07, -08 (2012).*
PA	Injection pressure shall not exceed maximum calculated to prevent new or propagation of fractures in the confining zone and shall not cause movement of injection or formation fluids into a protected aquifer. *40 C.F.R. §§ 144.51, 146.23 (2012).*
TX	Authorized pressure based on pressure test; regulations do not specify limit or formula. *16 Texas Admin. Code § 3.9 (2012).*
WY	Regulations do not specify limit or formula.[j]

Monitoring/reporting requirements

CO	Monthly reports are required, and are to include: types of chemicals used to treat injection water; the date of initial fluid injection for new injection wells; and the type and amount of fluids. Operators must record and report the volume of produced water; the volume of water injected into a Class II dedicated injection well; and the volume of water injected and produced in simultaneous injection wells. *2 Colo. Code Regs. §§ 404-1 (316A, 330) (2012).*

ND	The operator of an injection well shall meter or use an approved method to keep records and shall report monthly, including:
	• volume and nature (produced water, makeup water, etc.) of the fluid injected,
	• the injection pressure, and
	• such other information as may be required.
	Reports are required after completion or recompletion, or any remedial work that includes a detailed account of all work done, including the reason for the work; the date; the shots per foot, and size and depth of perforations; the quantity of sand, crude, chemical, or other materials employed in the operation; the size and type of tubing; the type and location of packer; the result of the packer pressure test; and other pertinent information that affects the status of the well. *N.D. Admin. Code 43-02-05-12 (2012).*
OH	Operator to monitor injection pressures and injection volumes daily, with average and maximum pressures and volumes compiled monthly and filed annually. (See also mechanical integrity testing.) *Ohio Admin. Code Ann. 1501:9-3-07 (2012).*[b]
PA	Permits are to specify monitoring requirements, including:
	(1) representative monitoring of the nature of injected fluids;
	(2) observation of injection pressure, flow rate, and cumulative volume (at specified frequencies depending on type of well and activity); and
	(3) recording of injection pressure, flow rate and cumulative volume at least monthly.
	Results are to be summarized in an annual report. *40 C.F.R. §§ 144.54, 146.23 (2012).*
TX	The operator shall monitor the injection pressure and injection rate of each disposal well monthly and report results annually. The operator shall report within 24 hours any significant pressure changes or other monitoring data indicating the presence of leaks in the well. *16 Texas Admin. Code § 3.9 (2012).*
WY	Operators shall report the type and source of the injected substance, the total amount injected, and the injected pressures and casing-tubing annulus pressure during injection. *Wyo. Code R. 055-000-004 § 10 (2012).*

Mechanical integrity testing

CO	Mechanical integrity tests are required initially and every 5 years to determine if there is:(1) a significant leak in the casing, tubing, or packer of the well, by pressure test, monthly pressure monitoring, or other approved test; and (2) any significant fluid movement into an underground source of drinking water through vertical channels adjacent to the wellbore, by tracer surveys, cement logs, temperature surveys, or other approved test. *2 Colo. Code Regs. § 404-1 (326) (2012).*
ND	Operator of a new injection well must demonstrate the mechanical integrity of the well prior to commencing operations, and at least once every 5 years. An injection well has mechanical integrity if: (1) there is no significant leak in the casing, tubing, or packer (demonstrated via a pressure test with liquid or gas, monitoring of positive annulus pressure following a valid pressure test, or a radioactive tracer survey); and (2) there is no significant fluid movement into an underground source of drinking water or an unauthorized zone through vertical channels adjacent to the injection bore (demonstrated via a log from which cement can be determined or well records demonstrating the presence of adequate cement to prevent such migration or a radioactive tracer survey, temperature log, or noise log). *N.D. Admin. Code 43-02-05-07 (2012).*
OH	Mechanical integrity to be documented by monitoring the annulus between the casing and tubing during injection of fluids at least monthly at a pressure sufficient to detect leaks, and reported annually. If such monitoring is not feasible, then once every 5 years the operator shall conduct mechanical integrity tests by pressure test, tracer surveys, noise logs; temperature surveys; or other approved test. *Ohio Admin. Code Ann. 1501:9-3-07 (2012).*[b]
PA	Mechanical integrity to be established prior to initial injection and tested once every 5 years:
	(1) must demonstrate absence of leaks by either monitoring of annulus pressure or pressure test with liquid or gas; (2) must demonstrate no significant fluid movement by results of a temperature or noise log; or cementing records demonstrating the presence of adequate cement to prevent such migration. *40 C.F.R. §§ 144.51,146.8, 146.23 (2012).*
TX	The mechanical integrity of a disposal well shall be evaluated by conducting pressure tests to determine whether the well tubing, packer, or casing have sufficient mechanical integrity to meet the prescribed performance standards. Each disposal well shall be tested for mechanical integrity prior to initial use, at least once every 5 years, and after every workover of the well. Mechanical integrity to be demonstrated by pressure test, or an approved alternate method such as tubing-casing annular pressure monitoring. The operator shall notify the Railroad Commission at least 48 hours prior to the testing. A complete record of all tests shall be filed within 30 days after the testing. *16 Texas Admin. Code § 3.9 (2012).*

WY	Mechanical integrity must be established by pressure testing at least once every 5 years to determine whether significant leaks are present in the casing, tubing, or packer. The initial mechanical integrity test for all disposal wells shall include one of the following tests to determine whether there are significant fluid movements in vertical channels adjacent to the wellbore:

- tracer surveys;
- cementing records with a cement bond log or other acceptable cement evaluation log;
- temperature surveys; or,
- any other test or combination of tests approved by EPA.

Operators must provide the Oil and Gas Conservation Commission the opportunity to witness all integrity tests. If not witnessed, the Operator is required to provide documentation of the test to the commission. If normal testing, surveys, or monitoring schedules provide inconclusive proof of mechanical reliability, the commission shall require that other appropriate logs or additional well tests be performed. *Wyo. Code R. 055-000-004 § 5 (2012).*

Approval prior to operation

CO	Yes. Each injection well must satisfactorily pass a mechanical integrity test prior to application approval and be approved prior to injection. *2 Colo. Code Regs. §§ 404-1 (325, 326) (2012).*
ND	Unclear. Prior to commencing operations, the operator of a new injection well must demonstrate the mechanical integrity of the well. Regulations do not specify approval. *N.D. Admin. Code 43-02-05-07 (2012).*
OH	Yes. Initial pressure testing of annulus between the tubing and the casing outside the tubing, under supervision of state, required prior to commencing injection. *Ohio Admin. Code Ann. 1501:9-3-05 (2012).*
PA	Yes, unless alternative schedule approved by EPA. *40 C.F.R. § 144.51 (2012).*
TX	Mechanical integrity of each disposal well shall be demonstrated prior to initial use. *16 Texas Admin. Code § 3.9 (2012).*
WY	Unclear. Testing required prior to operation. Regulations do not specify approval. *Wyo. Code R. 055-000-004 § 5 (2012).*

Plugging

CO	The operator must obtain approval from the Director of the plugging method prior to plugging and shall notify the Director of the estimated time and date the plugging operation of any well is to commence and identify the depth and thickness of all known sources of groundwater. Abandoned wells must be plugged in such a manner that oil, gas, water, or other substance shall be confined to the reservoir in which it originally occurred. Any cement plug shall be a minimum of 50 feet in length and shall extend a minimum of 50 feet above each zone to be protected. The top of the pipe must be sealed with either a cement plug and a screw cap, or cement plug and a steel plate welded in place or by other approved method, or marked with a permanent monument. All final reports of plugging and abandonment shall be submitted on a Well Abandonment Report and accompanied by a job log or cement verification report from the plugging contractor specifying the type of fluid used to fill the wellbore, type and slurry volume of American Petroleum Institute Class cement used, date of work, and depth the plugs were placed. *2 Colo. Code Regs. § 404-1 (319) (2012).*
ND	Well must be plugged with cement or other types of plugs, or both, in a manner that will not allow movement of fluids into an underground source of drinking water. The operator shall file a notice of intention to plug and obtain approval of the plugging method prior to the commencement of plugging operations. *N.D. Admin. Code 43-02-05-08 (2012).*
OH	Abandoned wells shall be plugged in such a manner that oil, gas, water, or other fluids shall be confined to the reservoir rock in which it occurs or originates. All Class II saltwater injection, enhanced recovery and Class III solution mining wells must be plugged with cement. Operators may use any method of emplacing the plugging material approved by the state but not limited to dump bailer, bullhead, pumping through tubing, casing, or drill pipe. Regulations specify intervals, including but not limited to:

- Wells must be plugged from total depth or a minimum of 50 feet below the base of the lowest reservoir rock penetrated to a minimum of 200 feet above the top of the lowest reservoir rock penetrated.
- From a minimum of 100 feet below to a minimum of 100 above the base of the surface casing.
- From a minimum of 100 feet below the grade level to 30 inches below grade level.

Each plugging operation shall be conducted under the supervision of an inspector. When plugging operations are not witnessed by an inspector, a plugging report is required. *Ohio Admin. Code Ann. 1501:9-11-03, -04, -08, -12 (2012).*

PA	Operator must submit plugging plan consistent with requirements. Well shall be plugged with cement to not allow the movement of fluids either into or between protected aquifers; allowed methods include (i) the Balance method; (ii) the Dump Bailer method; (iii) the Two-Plug method; or (iv) an approved comparable alternative. Report required within 60 days of plugging, certifying compliance or providing updated plan. *40 C.F.R. §§ 144.32, 144.51, 146.10 (2012).*
TX	Disposal wells shall be plugged upon abandonment. Disposal wells must meet plugging requirements for production wells, such as • insure that all formations bearing usable quality water, oil, gas, or geothermal resources are protected; • cement plugs shall be set as necessary to separate multiple usable quality water strata by placing the required plug at each depth as determined by the Texas Commission on Environmental Quality; • cement plugs shall be placed by the circulation or squeeze method through tubing or drill pipe; • additional requirements for cement and cementing; and • mud-laden fluid of at least 9-1/2 pounds per gallon with a minimum funnel viscosity of 40 seconds shall be placed in all portions of the well not filled with cement. The operator shall give the Railroad Commission advance notice of its intention to plug, and shall not commence the work until the proposed procedure has been approved. All cementing operations during plugging shall be performed under the direct supervision of the operator or his authorized representative. *16 Texas Admin. Code § 3.9, 3.14 (2012).*
WY	Wells must be plugged in a manner sufficient to properly protect all freshwater bearing formations and possible or probable oil or gas bearing formations. Plugging must be accomplished by the following: • All cement and additives shall consist of API class cement and additives; • Wells with production casing must be plugged by placing cement plugs of at least 100 foot length at least every 2,500 feet, in the base of the surface casing, and at least 100 feet inside the casing at the surface. If multiple casing strings are present, a minimum 100-foot plug must be placed in the annulus between each casing string at the outside casing shoe, and a minimum 100-foot plug in each annulus at the surface. • Regulations include other conditions of plugging, including requirements related to cement volume, perforations, and casing. Verbal approval to plug and abandon must be obtained prior to commencing actual plugging operations. When the well has been plugged, a notarized Subsequent Report of Abandonment accompanied by a job log or cement verification report from the plugging contractor specifying the type of fluid used to fill the wellbore, type of slurry volume of API Class cement used, date of work, and depth of plugs placed must be submitted to the Oil and Gas Conservation Commission. *Wyo. Code R. 055-000-003 § 18 (2012).*

Seismicity

CO	None identified. According to Colorado Oil and Gas Commission documents, its UIC permit review process was expanded in September 2011 to include a review for seismicity by the Colorado Geological Survey, and if historical seismicity is identified in the vicinity, the commission may require an operator to define the seismicity potential and the proximity to faults through geologic and geophysical data prior to any permit approval.[l]
ND	The application plan should depict faults, if known or suspected. All new injection wells shall be sited in such a fashion that they inject into a formation which has confining zones that are free of known open faults or fractures within the area of review. *N.D. Admin. Code 43-02-05-04, -05 (2012).*
OH	None identified.[b]
PA	Wells must be sited to inject into a formation which is separated from any protected aquifer by a confining zone that is free of known open faults or fractures within the area of review, which is either calculated or a minimum area within ¼ mile radius of the well. Applicant must identify faults if known or suspected within the area of review. *40 C.F.R. §§ 146.3, .6, .22, .24 (2012).*
TX	None identified.
WY	None identified.

Source: GAO analysis of state information.

aRequirements shown generally apply to new wells permitted after the early 1980s. Existing Class II wells, and new wells built in existing fields, were generally authorized by rule for up to five years from the effective date of the initial program, subject to conditions and requirements such as submission of inventory information. In Colorado, existing Class II enhanced recovery or hydrocarbon storage wells may be authorized by rule for the life of the well. Ohio requirements for annular disposal wells, enhanced recovery wells, and hydrocarbon storage wells may differ from those shown in this table.

bOhio Department of Natural Resources identified several reforms to its Class II deep injection well program, and proposed revisions to key regulations (1501:9-3-06, 9-3-07) including changes to application, testing, and monitoring requirements. See Ohio Department of Natural Resources, Preliminary Report on the Northstar 1 Class II Injection Well and the Seismic Events in the Youngstown, Ohio, Area (March 2012). In July 2012, the governor of Ohio signed an executive order determining that an emergency existed requiring the immediate adoption of the proposed rules. Rules filed as emergency rules remain in effect for 90 days, during which time Ohio Department of Natural Resources must go through the regular rule filing procedure. The emergency rules are undergoing public comment through August 31, 2012. Because the emergency rules are still subject to change through these processes, the table shows requirements as of June 2012, and does not reflect the emergency revisions.

cEPA implements the UIC program in Pennsylvania, so the table shows federal requirements applicable in the state of Pennsylvania.

dRequirements shown generally apply to new wells. Existing Class II wells were generally authorized by rule for up to five years from the effective date of the initial program, subject to conditions and requirements such as submission of inventory information. 40 C.F.R. § 144.21 (2011). EPA officials said the expectation was that existing Class II wells authorized by rule for 5 years and allowed to continue operations until permitted would eventually apply for and operate under a permit. In Pennsylvania, the effective date of the federal UIC program was June 25, 1984.

eSee generally 40 C.F.R. §§147.1951-1955, 144.1(f), pts. 144, 146 (2012).

fAccording to Texas state officials, however, every injection well permit includes a condition that states that "should it be determined that such injection fluid is not confined to the approved interval, then the permission given herein is suspended and the disposal operation must be stopped until the fluid migration from such interval is eliminated."

gEnhanced recovery or storage wells (e.g., wells used for injection of fluids into the producing formation) shall be cased with safe and adequate casing or tubing so as to prevent leakage, and shall be so set or cemented that damage will not be caused to oil, gas or freshwater resources. 2 Colo. Code Regs. § 404-1-404.

hSee also 40 C.F.R. § 147.305(d) (providing additional casing and cementing requirements that may be imposed by the EPA Regional Administrator).

iIn this table, "protected aquifer" refers to underground sources of drinking water; however, EPA UIC regulations define underground sources of drinking water as a subset of aquifers, namely an aquifer or its portion: (a)(1) Which supplies any public water system; or (2) which contains a sufficient quantity of groundwater to supply a public water system; and (i) currently supplies drinking water for human consumption; or (ii) contains fewer than 10,000 mg/l total dissolved solids; and (b) which is not an exempted aquifer. 40 C.F.R. § 144.3 (2011). EPA estimates there are approximately 1,000-2,000 exempted portions of aquifers.

jDisposal well permit applicants, are, however, required to submit evidence and data to support a state finding that the proposed disposal well will not initiate fractures through the overlying strata or confining zone which could enable the injection fluid or formation fluid to enter the freshwater strata. Wyo. Code R. 055-000-004 § 5 (2012). According to Wyoming state officials, pressure limits were administratively set on all injection wells existing before Nov. 22, 1982, and were set by orders on all Class II wells permitted after that time.

kOnly the initial tests are required for simultaneous injection wells. A simultaneous injection well is a well in which water produced from oil and gas producing zones is injected into a lower injection zone and such water production is not brought to the surface. 404-1-100.

lSee Colorado Oil and Gas Commission, COGCC Underground Injection Control and Seismicity in Colorado (Jan. 19, 2011).

Table 19: Selected State Requirements—Managing Air Emissions

Requirements related to hydrogen sulfide gas (H₂S)

CO	When well servicing operations take place in zones known to or reasonably expected to contain at or above 100 parts per million of H₂S, the operator shall file a H₂S drilling operations plan. Any gas analysis indicating the presence of H₂S shall be reported to the state and local government. *2 Colo. Code Regs. § 404-1(607) (2012).*
ND	Production facilities that emit sulfur compounds are subject to registration and reporting requirements. Anticipated H₂S content in produced gas from a proposed source of supply must be included in the application for permit to drill. The owner or operator of any oil or gas well production facility shall install equipment necessary to ensure that emissions comply with ambient air quality standards, including, but not limited to, H₂S. Each flare used for treating gas containing H₂S must be equipped and operated with an automatic ignitor or a continuous burning pilot that must be maintained in good working order. This is required even if the flare is used for emergency purposes only. Routine inspections and maintenance of tanks, hatches, compressors, vent lines, pressure relief valves, packing elements, and couplings must be conducted to minimize emissions from equipment used for gas containing H₂S. Tank hatches must hold a positive working pressure or must be repaired or replaced. *N.D. Admin. Code 33-15-20-01,-02, -04; 43-02-03-16 (2012).*
OH	A well that yields H₂S must be plugged with sulfate-resistant cement. In urbanized areas where there is a known occurrence of shallow gas or H₂S, drilling on air may not be permitted, fluid drilling shall be required. During drilling, the state inspector shall require converting to fluid drilling where there is an imminent threat to safety of the rig crew and/or the public. *Ohio Admin. Code Ann. 1501:9-11-07, -9-03 (2012).*
PA	An operator proposing to drill a well within a 1-mile radius of a well drilled to or through the same formation where H₂S has been found while drilling shall install monitoring equipment during drilling at the well site to detect the presence of H₂S in accordance with American Petroleum Institute publication API RP49, "Recommended Practices for Safe Drilling of Wells Containing H₂S." When H₂S is detected in concentrations of 20 parts per million or greater, the well shall be drilled in accordance with API RP49. An operator who operates a well in which H₂S is discovered in concentrations of 20 parts per million or greater shall operate the well in a way that presents no danger to human health or to the environment. When an operator discovers H₂S in concentrations of 20 parts per million or greater during the drilling of a well, the operator shall notify the state and identify the location of the well and the concentration of H₂S detected. The state will maintain a list of all notices that will be available to operators for their reference. *25 Pa. Code § 78.77 (2012).*
TX	H₂S emissions from source(s) must not exceed a net ground level concentration of 0.08 parts per million averaged over 30 minutes if the downwind concentration affects a residential, business or commercial property. Certain storage tanks must be posted with a warning sign on or within 50 feet of the facility; fencing is required when tanks are located inside towns or cities, or where tanks are exposed to the public; tanks are also subject to certain marker and compliance provisions. Operations where the 100 parts per million radius of exposure is greater than 50 feet are subject to warning and marker, security, and materials and equipment provisions. Certain other operations where the radius of exposure includes public areas or is greater than 3,000 feet are also subject to control and equipment safety provisions. Drilling and workover operations where the 100 parts per million radius of exposure is 50 feet or greater are subject to requirements related to protective breathing equipment; wind direction indicators installed; and automatic H₂S detection and alarm equipment. Drilling and workover operations where the 100 parts per million radius of exposure includes a public area or is 3,000 feet or greater are subject to additional provisions relating to protective breathing equipment; methods of igniting the gas in the event of an uncontrollable emergency; installation of a choke manifold, mud-gas separator, and flare line and provision of a suitable method for lighting the flare; secondary remote control of blowout prevention and choke equipment; drill stem testing of H₂S zones; certificates of compliance; pressure testing of blowout preventers and well control systems; and training. Operators may apply for exceptions. *30 Tex. Admin. Code § 112.31(2012); 16 Tex. Admin. Code § 3.36 (2012).*
WY	All flaring operations shall be conducted in a safe and workmanlike manner. If the gas stream is sour or venting would present a safety hazard, a constant flare igniter system or other state approved method to safely manage sour gas may be required. Venting of gas containing H₂S content in excess of 50 parts per million is generally not allowed. Venting does not include emissions associated with fugitive losses. State approval is required for venting of gas containing H₂S content in excess of 50 parts per million for specific job tasks in controlled environments, such as well repairs, pipeline purging, well failures, decommissioning of facilities, or where necessary as a safety measure where flaring would be dangerous due to the introduction of an ignition source at the work site or when the operation is conducted under the authority and regulations of the Department of Environmental Quality. *055-000-003 Code Wyo. R. § 39 (2012).*

Requirements related to venting and flaring

CO	Any gas escaping from the well during drilling operations shall be, so far as practicable, conducted to a safe distance from the well site and burned. The operator shall notify the local emergency dispatch as provided by the local governmental designee of any such flaring. Such notice shall be given prior to the flaring if it can be reasonably anticipated and, in all other cases, as soon as possible but in no more than 2 hours after it occurs. Unnecessary/excessive venting/flaring of natural gas produced from a well is prohibited. Except for gas flared or vented during an upset condition, well maintenance, well stimulation flowback, purging operations, or a productivity test, gas shall be flared or vented only after notice and approval. The notice shall estimate the volume and content of the gas and whether it contains more than 1 parts per million of H_2S. If necessary to protect the public health, safety, or welfare, the Director may require the flaring of gas.
	Gas flared, vented, or used on the lease shall be estimated based on a gas-oil ratio test or other approved equivalent test, and reported on Operator's Monthly Production Report. Flared gas shall be directed to a controlled flare or other combustion device operated as efficiently as possible to provide maximum reduction of air contaminants where practicable and without endangering the safety of the well site personnel and the public. Operators shall notify local emergency and government officials prior to flaring when it can be reasonably anticipated, or ASAP, but not more than 2 hours after it occurs.
	All salable quality gas shall be directed to the sales line as soon as practicable or shut in and conserved. Temporary flaring or venting shall be permitted as a safety measure during upset conditions and in accordance with all other applicable laws, rules, and regulations. In instances where green completion practices are not technically feasible or are not required, operators shall employ best management practices to reduce emissions. Such best management practices may include measures or actions, considering safety, to minimize the time period during which gases are emitted directly to the atmosphere, or monitoring and recording the volume and time period of such emissions. Such examples could include the flaring or venting of gas. *2 Colo. Code Regs. § 404-1(317, 912, 805) (2012).*
ND	Gas produced with crude oil from an oil well may be flared during a 1-year period from the date of first production from the well. Thereafter, flaring of gas from the well must cease and the well must be capped, connected to a gas gathering line, or equipped with an electrical generator that consumes at least 75 percent of the gas from the well. For a well operated in violation of this section, the producer shall pay royalties to royalty owners upon the value of the flared gas and shall also pay gross production tax on the flared gas. A producer may obtain an exemption from this section upon application and a showing that connection of the well to a natural gas gathering line is economically infeasible at the time of the application or in the foreseeable future or that a market for the gas is not available and that equipping the well with an electrical generator to produce electricity from gas is economically infeasible. Pending arrangements for disposition for some useful purpose, all vented casinghead gas shall be burned. Each flare shall be equipped with an automatic ignitor or a continuous burning pilot, unless waived by the state for good reason. *N.D. Cent. Code, § 38-08-06.4 (2012); N.D. Admin. Code 43-02-03-45 (2012).*
OH	All owners, lessees, or their agents, drilling for or producing crude oil or natural gas, shall use every reasonable precaution in accordance with the most approved methods of operation to stop and prevent waste of oil or gas, or both. Any well productive of natural gas in quantity sufficient to justify utilization shall be utilized or shut in within 10 days after completion. The owner of any well producing both oil and gas may burn such gas in flares when it is necessary to protect the health and safety of the public or when the gas is lawfully produced and there is no economic market at the well for the escaping gas. All gas vented to the atmosphere must be flared, with the exception of gas released by a properly functioning relief device and gas released by controlled venting for testing, blowing down, and cleaning out wells. Flares must be a minimum of 100 feet from the well, a minimum of 100 feet from oil production tanks and all other surface equipment, and 100 feet from existing inhabited structures and in a position so that any escaping oil or condensate cannot drain onto public roads or toward existing inhabited structures or other areas that could cause a safety hazard. In urbanized areas where flaring is expected, permittee shall notify local emergency response officials that such may occur. It is recommended that notice be provided if possible just prior to the expected flaring and/or immediately upon flare ignition. *Ohio Rev.Code Ann. § 1509.20 (2012). Ohio Admin. Code Ann. 1501:9-9-05, -03 (2012).*
PA	Excess gas encountered during drilling, completion, or stimulation shall be flared, captured, or diverted away from the drilling rig in a way that does not create a public health or safety hazard. *25 Pa. Code § 78.73 (2012).*
TX	Certain gas releases need not be flared, including: releases of gas that are not readily measured such as vapors from crude oil storage tanks; releases of gas from a well that must be unloaded or cleaned-up to atmospheric pressure (limited to 24 continuous hours or 72 hours in 1 month); releases of gas from a facility served by a gas gathering system, compression facility or gas plant (limited to 24 hours unless an exception is granted). Other authorized releases exceeding 24 hours shall be flared unless burning cannot be done safely. Such releases include: gas released for no more than 10 days after initial completion, recompletion in another field, or workover operations in the same field; hydrocarbon gas contained in the waste stream from a membrane unit or molecular sieve used to remove carbon dioxide, H_2S, or other contaminants from a gas

stream, provided that at least 85% of the hydrocarbon gas in the inlet gas stream is recovered and directed to a legal use; low pressure separator gas, not to exceed 15 mcfd of hydrocarbon gas per gas well or 50 mcfd of hydrocarbon gas per commission-designated oil lease or commingling point for commingled operations; releases resulting from cleaning a well of solids or fluids or both for more than 10 producing days following initial completion, recompletion in another field, or workover operations in the same field; releases resulting from unloading excess formation fluid buildup in a wellbore for periods in excess of 24 hours in one continuous event or 72 hours total in 1 calendar month; releases of volumes of low pressure gas that can be measured with devices routinely used in oil and gas exploration, development, and production operations and that are not directed by an operator to a gas gathering system, gas pipeline, or other marketing facility, or other purposes and uses authorized by law due to mechanical, physical, or economic impracticability; for casinghead gas only, releases associated with the unavailability of a gas pipeline or other marketing facility, or other purposes and uses authorized by law; and releases associated with avoiding curtailment of gas production which will result in a reduction of ultimate recovery from a gas well or oil reservoir. *16 Tex. Admin. Code § 3.32 (2012).*

WY | Venting or flaring during emergencies or upset conditions; well purging and evaluation tests; or production tests is not waste and is authorized. The state encourages employment of technologies that minimize or prevent venting and flaring during drilling and completion. Unless it is determined that waste is occurring, up to 60 MCF of gas per day is authorized to be vented or flared from individual oil wells. Venting or flaring is authorized either at the well or at a lease facility which serves several wells. An Owner/Operator must apply for retroactive or prospective venting or flaring authorization in other circumstances. Authorization may be granted if the venting or flaring does not constitute waste.*055-000-003 Code Wyo. R. § 39 (2012).*

Source: GAO analysis of state information.

Appendix X: Crosswalk between Selected Requirements from EPA, States, and Federal Lands

Table 20 is intended to show representative areas of regulation, focused on substantive requirements specific to oil and gas wells. The table includes EPA's environmental and public health requirements, requirements from the six states included in our review, and additional requirements that apply for the development of federally-owned mineral resources. Other activities at oil and gas well sites may also be subject to federal or state regulation.

Table 20: Crosswalk between Selected Requirements from EPA, Six States, and Federal Minerals

Area of regulation	EPA environmental and public health requirements	Requirements of six states reviewed	Additional requirements for federal minerals
Siting and site preparation			
Comprehensive environmental assessment prior to drilling	No[a]	[b]	Yes
Identification or testing of water wells prior to drilling of production wells	No	1 of 6 (Wyoming) [identification alone] 2 of 6 (Colorado, Ohio) [identification and testing][c]	Yes – identification No – testing
Required setbacks from water sources	No[d]	5 of 6 (Colorado, North Dakota, Pennsylvania, Ohio, Wyoming)	Yes
Erosion control, site preparation, surface disturbance minimization, and stormwater management	Effectively no[e]	6 of 6 [any] 4 of 6 (Colorado, North Dakota, Pennsylvania, Wyoming) [stormwater permitting][f]	Yes
Drilling, casing, and cementing			
Requirements relating to cementing/casing plans	No[g]	6 of 6[h]	Yes
Prescribed placement of surface casing relative to groundwater zones	No[g]	6 of 6	Yes
Prescribed cementation techniques for surface casing	No[g]	6 of 6	No; instead the Bureau of Land Management (BLM) has performance standards
Requirement for cement waiting period and/or integrity tests	No[g]	6 of 6	Yes
Blowout preventer[i] requirements	No[g]	5 of 6[j] (Colorado, North Dakota, Pennsylvania, Texas, Wyoming)	Yes
Hydraulic fracturing			
Prior authorization/notice/inspection requirements	No	4 of 6 (Colorado, Ohio, Pennsylvania, Wyoming)	Not currently, but in BLM proposed rule
Requirements to disclose information on fracturing fluids	No[k]	6 of 6	Not currently, but in BLM proposed rule

Area of regulation	EPA environmental and public health requirements	Requirements of six states reviewed	Additional requirements for federal minerals
Pressure monitoring, testing, limitations or other mechanical integrity requirements specific to hydraulic fracturing	No	4 of 6 (Colorado, North Dakota, Ohio, Wyoming)	Not currently, but in BLM proposed rule
Well plugging			
Requirements for notification, plugging plan or method, witnessing, and reporting	No[l]	6 of 6	Yes
Programs to plug wells that are not properly plugged and have been abandoned	No	6 of 6	Yes[m]
Site reclamation			
Requirements for backfilling, regrading, recontouring, and alleviating compaction of soil	No	6 of 6	Yes
Revegetation requirements	No	5 of 6 (Colorado, North Dakota, Ohio, Pennsylvania, Wyoming)	Yes
Waste management			
Options for waste disposal:			
Underground injection	Yes (Safe Drinking Water Act)	5 states have their own requirements (Colorado, North Dakota, Ohio, Texas, Wyoming); EPA implements the program in Pennsylvania	In the permit application, operators must describe the final disposal of waste materials.
Direct discharge to surface water	Yes (Clean Water Act—certain discharges prohibited, others subject to conditions and permits)	Surface discharges are allowed in certain cases in 3 western states (Colorado, Texas, Wyoming)	In the permit application, operators must describe the final disposal of waste materials.
Requirements for discharge to publicly-owned treatment works (POTWs) or Centralized Waste Treatment Facilities (CWT)	Pretreatment standards for shale gas wastewater under development (Clean Water Act)	Disposal at POTWs is an option in two states[n] (Ohio, Pennsylvania) Disposal at CWT facilities is an option in 3 states (Colorado, Pennsylvania, Wyoming)	In the permit application, operators must describe the final disposal of waste materials.
Recycling or other reuse	Yes (Clean Water Act—certain produced water discharges)	6 of 6 states allow recycling or other reuse[o]	In the permit application, operators must describe the final disposal of waste materials.
Solid waste disposal	Effectively no[p]	Yes	In the permit application, operators must describe the final disposal of waste materials.

Area of regulation	EPA environmental and public health requirements	Requirements of six states reviewed	Additional requirements for federal minerals
Hazardous waste disposal	Effectively no[q]	No	In the permit application, operators must describe the final disposal of waste materials.
Pit siting requirements (with regard to sensitive areas)	No[r]	6 of 6	No specific requirements but pits must be approved
Pit lining requirements	No	5 of 6[s] (Colorado, North Dakota, Pennsylvania, Texas, Wyoming)	Not currently, but BLM proposed rule would require liners for pits used to store hydraulic fracturing flowback fluids.
Pit closure requirements	No	6 of 6	Yes
Managing air emissions			
Requirements for criteria pollutants	Certain Clean Air Act provisions apply	5 of 6 states have permitting or registration programs (Colorado, North Dakota, Ohio, Texas, Wyoming)[t]	No
Requirements for hazardous air pollutants	Certain Clean Air Act provisions apply	State permitting or registration programs may address hazardous air pollutants	No
Requirements related to hydrogen sulfide gas	No specific requirements[u]	6 of 6	Yes
Requirements related to flaring	Under new New Source Performance Standards, most hydraulically fractured gas wells must do green completions	6 of 6	Yes

Sources: GAO analysis of federal and state laws and regulations.

[a]Under the National Environmental Policy Act (NEPA), federal agencies must assess the effects of major federal actions—those they propose to carry out or to permit—that significantly affect the environment. Many Environmental Protection Agency (EPA) activities relevant here are exempt from NEPA's procedural requirements by statute or recognition by courts that EPA procedures or environmental reviews under enabling legislation are functionally equivalent to the NEPA process. See 63 Fed. Reg. 58045 (Oct. 29, 1998).

[b]We did not specifically analyze state requirements in this area. However, when asked whether they had a comprehensive environmental assessment process prior to drilling, officials from Ohio, Texas, Pennsylvania, and Wyoming said they did not. Officials in North Dakota said that an environmental assessment is required for drilling on state lands and officials in Colorado said that a location assessment is required, which includes assessing transportation access, future reclamation plans, and determination of whether the proposed location is within a sensitive wildlife habitat.

[c]Testing requirement applies only to certain wells—certain wells near proposed coalbed methane wells in Colorado and wells proposed for urbanized areas or in the vicinity of horizontal wells in Ohio. Pennsylvania does not require operators to identify or test nearby water wells, but state law incentivizes operators to do so by establishing a rebuttable presumption that operators are liable for changes in water quality of certain wells after drilling.

[d]There are no federal requirements regarding setbacks, but under Section 404 of the Clean Water Act, a permit from the U.S. Army Corps of Engineers is required to fill waters of the United States, such as wetlands.

[e]Oil and gas well sites are only required to get permits for stormwater discharges if the facility has had a discharge of contaminated stormwater that includes a reportable quantity of a pollutant or contributes to the violation of a water quality standard, rather than prior to commencing construction or causing discharges.

[f]Ohio and Texas state regulations address stormwater in other ways.

[g]Generally federal environmental laws do not have drilling, cementing, or casing requirements related to drilling production wells. However, according to EPA officials, if the well is to be hydraulically fractured with diesel fuel, it is subject to regulation as a Class II well under the underground injection control (UIC) program authorized by the Safe Drinking Water Act, and be subject to cementing and casing requirements. See 40 C.F.R. §§ 144.52 and 146.22. In May 2012, EPA published draft guidance on how its UIC permit writers should address hydraulic fracturing with diesel in the context of the Class II UIC program. To date, however, EPA officials are unaware of any wells that were regulated in this way.

[h]Colorado, North Dakota, Ohio, Pennsylvania, and Wyoming require cementing/casing plans. Texas requires cementing/casing plans if an operator proposes a method of freshwater protection other than those prescribed by state regulations.

[i]Blowout preventers are devices placed on wells to help maintain control over pressures in the well and prevent the well from spewing oil, gas, or other formation fluids in the case of a blowout.

[j]North Dakota, Texas, and Wyoming require blowout preventers; Colorado and Pennsylvania require blowout preventers in certain circumstances.

[k]Under TSCA, to the extent a hydraulic fracturing fluid is a chemical substance or mixture, manufacturers (including importers), processors, and distributors of such fluids generally would be subject to applicable reporting requirements. Generally, well site operators would not be subject to any such applicable TSCA reporting requirements.

[l]Generally federal environmental laws do not have requirements related to well plugging. However, according to EPA officials, if the well is to be hydraulically fractured with diesel fuel, it is subject to regulation as a Class II UIC well under the SDWA UIC program, as discussed in table note g above.

[m]According to BLM officials, BLM periodically conducts orphan and idle well operations.

[n]Disposal at a POTW is currently available in Pennsylvania and Ohio. Discharges may be authorized from a POTW in Pennsylvania if preceded by treatment at a CWT. The Ohio Environmental Protection Agency recently prohibited disposal at a POTW in the state, but an administrative review commission removed the prohibition as beyond the agency's authority. The Ohio Department of Natural Resources may take separate action to prohibit the practice. We are also aware that the city of Forth Worth, Texas had a pilot program within the last several years under which it accepted flowback for disposal through its POTW, but current information suggests that the city is no longer accepting flowback water.

[o]For example, four states allow operators to reuse certain types of fluid waste for road applications.

[p]The existing federal regulations under RCRA solid waste provisions apply to nonhazardous waste disposal facilities and practices, including those involving oil and gas wastes, and prohibit open dumping of solid waste. However, EPA has a limited role in the enforcement of RCRA solid waste provisions.

[q]Per EPA's 1988 regulatory determination, oil and gas exploration and production wastes—including wastes originating in the well or associated field operations—are not regulated as hazardous. Small amounts of other hazardous waste may be at well sites (such as discarded, unused hydraulic fracturing fluids) but we could not identify any instances where these wastes were available in high enough quantities to trigger RCRA requirements.

[r]Under Section 404 of the Clean Water Act, a permit from the U.S. Army Corps of Engineers is required to fill waters of the United States, such as wetlands.

[s]Colorado, North Dakota, Pennsylvania, and Wyoming have specific pit lining requirements. Texas regulations provide for pit lining requirements to be addressed in permits.

[t]In addition, Pennsylvania is in the process of developing an inventory for oil and gas emissions information.

[u]Although there are no specific requirements, owners and operators are subject to the Clean Air Act general duty clause to take steps to prevent accidental releases of listed and other substances to the air; these include hydrogen sulfide.

Appendix XI: Comments from the Department of Agriculture

USDA United States Department of Agriculture   Forest Service   Washington Office   1400 Independence Avenue, SW Washington, DC 20250

File Code: 1420

Date: AUG 1 5 2012

Mr. David Trimble
Director, Natural Resources and Environment
U.S. Government Accountability Office
441 G Street, N.W.
Washington, DC 20548

Dear Mr. Trimble:

Thank you for the opportunity to review and provide comment on the draft U.S. Government Accountability Office (GAO) report on "Unconventional Oil and Gas Development: Key Environmental and Public Health Requirements" (GAO-12-874). The Forest Service has reviewed the report and generally concurs with the information provided in the report and noted that there are no specific recommendations or findings.

Relative to the description of challenges identified by the various agencies, page 70 of the report addresses the Bureau of Land Management challenge of hiring and retaining qualified staff in areas with very active oil and gas development. The Forest Service shares that challenge, especially on the Dakota Prairie Grassland.

We compliment GAO in the use of graphics in the report to facilitate visual display of operations or processes, which are very helpful in understanding quite technical information.

If you have any questions, please contact Thelma Strong, Acting Chief Financial Officer, at 202-205-1321 or tstrong@fs.fed.us.

Sincerely,

THOMAS L. TIDWELL
Chief

Caring for the Land and Serving People   Printed on Recycled Paper

Appendix XII: Comments from the Department of the Interior

United States Department of the Interior

OFFICE OF THE SECRETARY
Washington, DC 20240

AUG 2 0 2012

Mr. David C. Trimble
Director, Natural Resources and Environment
U.S. Government Accountability Office
441 G Street, N.W.
Washington, D.C. 20548

Dear Mr. Trimble:

On July 25, 2012, the Government Accountability Office (GAO) issued a draft report on hydraulic fracturing entitled, *UNCONVENTIONAL OIL AND GAS DEVELOPMENT: Key Environmental and Health Requirements* (GAO-12-874). The GAO made no recommendations to the Department of the Interior (Department). However, the Department offers comments on certain findings and conclusions contained in the report. We believe these additions are necessary to create a fuller public understanding of Departmental initiatives that are otherwise only briefly mentioned in the report.

Proposed Hydraulic Fracturing Rule

The Department appreciates the GAO's recognition of the Bureau of Land Management's (BLM's) proposed rule on hydraulic fracturing, which the report describes briefly (p. 65). This proposed rule is a significant advance toward responsible development of unconventional oil and gas resources on the public lands, with the potential to increase public confidence in its regulation. The BLM's proposed rule focuses on disclosure of chemicals used in the process, well integrity, and management of flowback fluids. We believe that the report would benefit from greater attention to this proposed rule, and therefore request that the GAO address additional information in its discussion of the proposed rule, as detailed below.

Hydraulic Fracturing Rule Characterization
Secretary Salazar has placed significant policy and program emphasis on the establishment of a new regulatory framework to assist industry and the BLM in managing hydraulic fracturing on Federal and Tribal lands among the many states in which BLM oversees leasing of the Federal subsurface estate. The BLM currently administers Federal and Indian oil and gas leases across 34 states. BLM regulations on hydraulic fracturing will apply consistently to all leases managed by the BLM, whereas state management of such leases may differ by jurisdiction. Consistent standards governing BLM-managed subsurface will help the Bureau achieve sensible, responsible stewardship of the public estate.

There are three main aspects of the BLM's proposed rule: 1) the requirement for operators to publicly disclose all chemicals used during hydraulic fracturing operations, while protecting trade secrets; 2) the requirement to confirm wellbore integrity before and during the hydraulic fracturing operation; and 3) the requirement to properly manage waters that flow back to the surface from hydraulic fracturing operations. The Department believes this rule will result in

2

significant improvements for the BLM's management of unconventional oil and gas development on Federal and Indian lands, and increase public confidence in this industry practice.

BLM Mission and Framework
The GAO report should more fully articulate distinctions between BLM and state land management responsibilities. We suggest adding additional language to the report's description of the proposed rule on p. 65 to include the following information: Pursuant to the Federal Land Policy and Management Act (FLPMA) and other applicable laws, the BLM has a mission to manage the public lands for multiple use and sustained yield, in the national interest. This imposes stewardship responsibilities that may differ, substantially in many cases, from state authorities and interests addressed in the report. State lands are often managed solely or primarily for revenue, and state provisions governing oil and gas activities on state-administered lands may not address conservation needs and natural resource values, such as air, soil, water, wildlife, recreation, and cultural and scenic landscapes, in a manner that would satisfy FLPMA's multiple-use directive. The BLM's multiple-use mission is reflected in the BLM's proposed rule on hydraulic fracturing, which is intended to improve stewardship and operational efficiency by establishing a uniform set of standards for hydraulic fracturing practices across the public and Indian lands.

Cement Bond Logs and Surface Casing
The Department also requests that the GAO report more specifically describe the provisions of the proposed rule relating to well bore integrity and the requirements for cement bond logs and surface casing. These elements of unconventional gas recovery are of primary importance to surrounding landowners, stakeholders, and the public, and the Department believes that they merit further explanation in the report. The BLM's proposed regulations would require that an operator take the following steps to ensure well bore integrity: 1) run a string of steel pipes (surface and intermediate casing) across all usable water zones; 2) cement the steel pipes to the rock formation in a manner that seals all spaces behind the pipe; and 3) use available technology (e.g., cement bond logs) to verify that there is a secure cement seal between the steel pipes and the rock formation to avoid fluid migration outside the pipes. Finally, at the conclusion of drilling and before hydraulic fracturing takes place, the proposed rule would require an operator to test the steel pipes to ensure that the steel pipe construction is suitable to withstand the force and pressure to be applied by the hydraulic fracturing activity. The BLM's requirements in the proposed rule are consistent with the American Petroleum Institute's published guidelines on industry best practices for mitigating environmental impacts that may be associated with hydraulic fracturing operations. The BLM will consider public comments on this proposal.

Aquifer Protection
The Department also requests that the GAO more specifically articulate the provisions of the proposed rule relating to the protection of aquifers. The provisions relating to well bore integrity through casing and cementing isolation, described above, provide important protection for aquifers. Additionally, the proposed rule includes requirements for operators to submit the length and height of the hydraulic fractures, both before and after the actual fracturing, so that the BLM can ensure the safety and zonal isolation integrity of aquifers in the vicinity of the oil

3

and gas reserves. Overall, the BLM has taken measures to protect aquifers through the proposed rule's commitment to wellbore integrity, mechanical integrity, and the management of the flowback water (as discussed below).

Flowback

The Department also requests that the report include a description of provisions in the proposed rule regarding the collection of information about and management of disposal of flowback water. Flowback water (fluids recovered after fracturing), if mishandled, can pose a risk to water supply. The BLM's proposed rule therefore would require that an operator store flowback water in a steel tank and/or a lined pit. The proposed rule would also mandate that operators estimate the volume of flowback water and report how they will handle it. The proposed rule would further specify that operators must comply with the BLM's *Onshore Oil and Gas Order Number 7, Disposal of Produced Water*, when managing flowback fluids associated with hydraulic fracturing operations. The BLM believes that the above steps are necessary to protect water and natural resources at the surface from contamination by hydraulic fracturing. Further, these steps are consistent with industry practice and guidelines published by the American Petroleum Institute. Finally, in the proposed rule the BLM reserves the right to place additional restrictions and issue additional guidelines on the handling of flowback water, as appropriate.

Inspection and Enforcement Workforce Strategy Report

The GAO Report also mentions difficulties faced by the BLM in hiring and retaining skilled technical staff (pp.70-71). A BLM team consisting of field, state, and headquarters staff has reviewed the ongoing issues related to hiring, training, and retaining qualified Petroleum Engineers and Petroleum Engineering Technicians. The BLM is in the process of developing a report that will provide current information related to these issues and make recommendations for improvements. We request that the GAO report acknowledge this initiative.

Thank you for the opportunity to comment on this report and its analysis of the important topic of hydraulic fracturing. In addition to the above comments, we also enclose a number of technical comments. If you have any questions, please contact Nicholas Douglas, Senior Policy Advisor, at 202-912-7311 or LaVanna Stevenson-Harris, BLM Audit Liaison Officer, at 202-912-7077.

Sincerely,

Ned Farquhar

Ned Farquhar
Deputy Assistant Secretary
Land and Minerals Management

Enclosure

Appendix XIII: GAO Contact and Staff Acknowledgments

GAO Contact	David C. Trimble, (202) 512-3841 or trimbled@gao.gov
Staff Acknowledgments	In addition to the individual named above, Barbara Patterson, Assistant Director; Elizabeth Beardsley; David Bieler; Antoinette Capaccio; Cindy Gilbert; Armetha Liles; Alison O'Neill; and Janice Poling made key contributions to this report.

GAO's Mission	The Government Accountability Office, the audit, evaluation, and investigative arm of Congress, exists to support Congress in meeting its constitutional responsibilities and to help improve the performance and accountability of the federal government for the American people. GAO examines the use of public funds; evaluates federal programs and policies; and provides analyses, recommendations, and other assistance to help Congress make informed oversight, policy, and funding decisions. GAO's commitment to good government is reflected in its core values of accountability, integrity, and reliability.
Obtaining Copies of GAO Reports and Testimony	The fastest and easiest way to obtain copies of GAO documents at no cost is through GAO's website (www.gao.gov). Each weekday afternoon, GAO posts on its website newly released reports, testimony, and correspondence. To have GAO e-mail you a list of newly posted products, go to www.gao.gov and select "E-mail Updates."
Order by Phone	The price of each GAO publication reflects GAO's actual cost of production and distribution and depends on the number of pages in the publication and whether the publication is printed in color or black and white. Pricing and ordering information is posted on GAO's website, http://www.gao.gov/ordering.htm.
	Place orders by calling (202) 512-6000, toll free (866) 801-7077, or TDD (202) 512-2537.
	Orders may be paid for using American Express, Discover Card, MasterCard, Visa, check, or money order. Call for additional information.
Connect with GAO	Connect with GAO on Facebook, Flickr, Twitter, and YouTube. Subscribe to our RSS Feeds or E-mail Updates. Listen to our Podcasts. Visit GAO on the web at www.gao.gov.
To Report Fraud, Waste, and Abuse in Federal Programs	Contact: Website: www.gao.gov/fraudnet/fraudnet.htm E-mail: fraudnet@gao.gov Automated answering system: (800) 424-5454 or (202) 512-7470
Congressional Relations	Katherine Siggerud, Managing Director, siggerudk@gao.gov, (202) 512-4400, U.S. Government Accountability Office, 441 G Street NW, Room 7125, Washington, DC 20548
Public Affairs	Chuck Young, Managing Director, youngc1@gao.gov, (202) 512-4800 U.S. Government Accountability Office, 441 G Street NW, Room 7149 Washington, DC 20548

Please Print on Recycled Paper.

www.ingramcontent.com/pod-product-compliance
Lightning Source LLC
Chambersburg PA
CBHW081112170526
45165CB00008B/2417